T0305756

HIGHER MATHEMATICS FOR ENGINEERING AND TECHNOLOGY

Problems and Solutions

HIGHER MATHEMATICS FOR ENGINEERING AND TECHNOLOGY

Problems and Solutions

Mahir M. Sabzaliev, PhD
Ilhama M. Sabzalieva, PhD

Apple Academic Press Inc.
3333 Mistwell Crescent
Oakville, ON L6L 0A2
Canada

Apple Academic Press Inc.
9 Spinnaker Way
Waretown, NJ 08758
USA

Printed in the United States of America on acid-free paper

International Standard Book Number-13: 978-1-77188-642-0 (Hardcover)

International Standard Book Number-13: 978-0-203-73013-3 (eBook)

Library and Archives Canada Cataloguing in Publication

Sabzaliev, Mahir M., author
Higher mathematics for engineering and technology : problems and solutions / Mahir M. Sabzaliev, PhD, Ilhama M. Sabzalieva, PhD.

Includes bibliographical references and index.
Issued in print and electronic formats.
ISBN 978-1-77188-642-0 (hardcover).--ISBN 978-0-203-73013-3 (PDF)

1. Engineering mathematics. I. Sabzaliev, Ilhama M., author II. Title.

TA330.S23 2018 620.001'51 C2018-900583-1 C2018-900584-X

CIP data on file with US Library of Congress

Apple Academic Press also publishes its books in a variety of electronic formats. Some content that appears in print may not be available in electronic format. For information about Apple Academic Press products, visit our website at **www.appleacademicpress.com** and the CRC Press website at **www.crcpress.com**

ABOUT THE AUTHORS

Mahir M. Sabzaliev

Mahir M. Sabzaliev is head of Higher Mathematics and Technical Sciences chair at Baku Business University. He is also a professor of general and applied mathematics at Azerbaijan University of Oil and Industry, Baku, Azerbaijan, where he was a head of the higher mathematics chair in 2011–2015. He is a member of the International Teachers Training Academy of Science. He has authored over 100 published scientific works, including 30 educational works and scientific-methodical aids. He has given many talks at international conferences. His papers were published in several well-known journals, including *Doklady Academy of Sciences of SSSR, Doklady of Russian Academy of Sciences, Differential Equations (Differentsial'nye Uravneniya),* and *Uspekhi Matematicheskikh Nauk,* among others.

Dr. Sabzaliev graduated from Azerbaijan State Pedagogical University with an honors diploma in mathematics. He worked as a teacher of mathematics in a secondary school and subsequently enrolled as a full-time post graduate student and earned the candidate of physical-mathematical sciences degree. In 2013, he earned a PhD in mathematics.

Ilhama M. Sabzalieva

Ilhama M. Sabzalieva, PhD, is an associate professor of general and applied mathematics and also department chair at Azerbaijan University of Oil and Industry, Baku, Azerbaijan. She has authored over 40 scientific works, including 10 educational works and scientific-methodical aids. She has authored more than 40 scientific works and has prepared educational supplies and scientific-methodical aids. She has attended several international conferences and given talks. She has also published papers in journals such as *Doklady of Russian Academy of Sciences* and *Differentsial'nye Uravneniya*. Dr. Sabzalieva graduated from Azerbaijan State University of Oil and Industry with an honors diploma and also earned the candidate of physical-mathematical sciences degree.

CONTENTS

PREFACE

This book was prepared and written based on the long-term pedagogical experience of the authors at Azerbaijan State University of Oil and Industry.

The problems that cover all themes of mathematics on engineering-technical specialties of higher technical schools have been gathered in this volume, which consists of three parts. The volume contains sections on "elements of linear algebra and analytic geometry," "differential calculus of a function of one variable," and "higher algebra elements."

In this book, on every theme we present short theoretical materials and then give problems to be solved in class or independently at home, along with their answers. On each theme we give the solution of some typical, relatively difficult problems and guidelines for solving them.

In the case of when students will be working out the problems to be solved independently, we have taken into account the problems' similarity with the problems to be solved in class, and we stress the development of the self-dependent thinking ability of students.

The problems marked by "*" are relatively difficult and are intended for students who want to work independently.

This book is intended for bachelor students of engineering-technical specialties of schools of higher education and will also be a good resource for those beginning in various engineering and technical fields. The book will also be valuable to mathematics faculty, holders of master's degrees, engineering staff, and others.

CHAPTER 1

ELEMENTS OF LINEAR ALGEBRA AND ANALYTIC GEOMETRY

CONTENTS

ABSTRACT

In this chapter, we give brief theoretical materials on a matrix, determinant, operations on matrixes, finding of inverse matrix, calculating the rank of a matrix, a system of linear equations and methods for solving them, basis vectors, scalar, vectorial and mixed products of vectors, straight-line theoretical equations on a plane and space, parabola, and 217 problems.

1.1 MATRICES AND OPERATIONS

Under a matrix, one understands a table in the form of a rectangle made of numbers. We can write a matrix with row m and column n in the form:

$$A = \begin{pmatrix} a_{11} & a_{12} & . & . & . & a_{1n} \\ a_{21} & a_{22} & . & . & . & a_{2n} \\ . & . & . & . & . & . \\ a_{m1} & a_{m2} & . & . & . & a_{mn} \end{pmatrix}$$

The members m and n are the sizes of the matrix. A matrix with row m and column n is said to be $m \times n$ dimensional matrix. For $m \neq n$ A is called a rectangle, for $m = n$, A is said to be an n-th order square matrix. Sometimes, $m \times n$ dimensional matrix is written in brief as follows:

$$A = \left(a_{ij}\right); \quad i = 1, 2, ..., m; \quad j = 1, 2, ..., n.$$

The entries composing a matrix are called its elements. The notation a_{ij} shows the elements standing in the intersection of the i-th row and j-th column of the matrix. Sometimes, the elements of the matrix may be algebraic expressions, functions, etc.

Any matrix may be multiplied by a number, the same dimensional matrices may be put together or subtracted, when the number of the columns of the first matrix equals the number of rows of the second matrix, the first matrix may be multiplied by the second matrix.

In order to multiply the matrix by any number, all elements of this matrix must be multiplied by this number. When λ is an arbitrary number, we can write it as follows $A \cdot \lambda = \left(a_{ij} \cdot \lambda\right); \quad i = 1,\ 2,\ ...\ ,\ m; \quad j = 1,\ 2,\ ...\ ,\ n.$

In order to put together two matrices, appropriate elements of these matrices should be put together.

If we denote the sum of the matrices, $A = (a_{ij})$; $B = (a_{ij})$; $i = 1, 2, \ldots, m$; $j = 1, 2, \ldots, n$
by

$$C = (c_{ij}); \text{ i} = 1, 2, \ldots, m; \quad j = 1, 2, \ldots, n,$$

we can write the rule of addition of matrices in the form

$$c_{ij} = a_{ij} + b_{ij}.$$

When subtracting the matrices, appropriate matrices of the second matrix are subtracted from the elements of the first matrix.

The operations of multiplication by a number, and addition of matrices have the following properties:

1. When λ and μ are any numbers, A is an arbitrary matrix, then
$$\lambda\,(\mu\, A) = \mu\,(\lambda\, A) = (\lambda\, \mu)\, A;$$
2. When λ is an arbitrary number, A and B are any same-dimensional matrices, then
$$\lambda\,(A + B) = \lambda\, A + \lambda\, B;$$
3. When λ and μ are any numbers, A is an arbitrary matrix, then
$$(\lambda + \mu)\, A = \lambda\, A + \mu\, A;$$
4. When A and B are the same-dimensional matrices, then
$$A + B = B + A;$$
5. When A, B, and C are any same-dimensional matrices, then
$$(A + B) + C = A + (B + C)$$

Now we give the rule of multiplication of matrices. Suppose that the number of the columns of the matrix A is equal to the number of rows of the matrix B. For example, A is $m \times n$ dimensional, B is $n \times p$ dimensional:

$$A = (a_{ij}), \;\; i = 1, 2, \ldots, m; \quad j = 1, 2, \ldots, n,$$
$$B = (b_{ij}), \;\; i = 1, 2, \ldots, n; \quad j = 1, 2, \ldots, p\,.$$

At first, we note that $A \cdot B = C$ matrix will be $m \times p$ dimensional. For finding the element c_{ij} standing in the intersection the i-th row and the j-th column of the matrix $C = (c_{ij})$, we multiply the i-th elements of A by the appropriate elements of the j-th column of B and put together the obtained products. We write this rule by formula as follows:

$$c_{ij} = a_{i1} b_{1j} + a_{i2} b_{2j} + \dots + a_{in} b_{nj}; \quad i = 1, 2, \dots, m; \quad j = 1, 2, \dots, p$$

Generally speaking, law of permutation is not true for multiplication of matrices.

Problems to be solved in auditorium

Problem 1. For $A = \begin{pmatrix} 2 & 1 & -1 \\ 0 & 1 & -4 \end{pmatrix}$, $B = \begin{pmatrix} -2 & 1 & 0 \\ -3 & 2 & 2 \end{pmatrix}$ find the matrix:

1) $3A + 2B$; 2) $2A - 4B$.

Answer: 1) $\begin{pmatrix} 2 & 5 & -3 \\ -6 & 7 & -8 \end{pmatrix}$; 2) $\begin{pmatrix} 12 & -2 & -2 \\ 12 & -6 & -16 \end{pmatrix}$.

Problem 2. Calculate:

1) $\begin{pmatrix} 2 & -3 \\ 4 & -6 \end{pmatrix} \cdot \begin{pmatrix} 9 & -6 \\ 6 & -4 \end{pmatrix}$; 2) $\begin{pmatrix} 1 & -3 & 2 \\ 3 & -4 & 1 \\ 2 & -5 & 3 \end{pmatrix} \cdot \begin{pmatrix} 2 & 5 & 6 \\ 1 & 2 & 5 \\ 1 & 3 & 2 \end{pmatrix}$; 3) $\begin{pmatrix} 1 & 2 \\ 3 & 4 \end{pmatrix}^2$.

Answer: 1) $\begin{pmatrix} 0 & 0 \\ 0 & 0 \end{pmatrix}$; 2) $\begin{pmatrix} 1 & 5 & -5 \\ 3 & 10 & 0 \\ 2 & 9 & -7 \end{pmatrix}$; 3) $\begin{pmatrix} 7 & 10 \\ 15 & 22 \end{pmatrix}$.

Problem 3. Calculate the product of matrices:

1) $\begin{pmatrix} 5 & 0 & 2 & 3 \\ 4 & 1 & 5 & 3 \\ 3 & 1 & -1 & 2 \end{pmatrix} \cdot \begin{pmatrix} 6 \\ -2 \\ 7 \\ 4 \end{pmatrix}$; 2) $\begin{pmatrix} 1 & 0 & -2 \\ 2 & 3 & -1 \end{pmatrix} \cdot \begin{pmatrix} 3 & -2 & 4 & 0 \\ 2 & -1 & 5 & -4 \\ 1 & 0 & -6 & -3 \end{pmatrix}$;

$$3)\ \begin{pmatrix} 4 & 0 & -2 & 3 & 1 \end{pmatrix}\cdot\begin{pmatrix} 3 \\ 1 \\ -1 \\ 5 \\ 2 \end{pmatrix};\quad 4)\ \begin{pmatrix} 1 & 3 & 5 \\ 2 & 4 & 6 \end{pmatrix}\cdot\begin{pmatrix} 1 & 2 \\ 0 & 1 \\ -1 & 0 \end{pmatrix}.$$

Solution of 1): As the size of the first matrix is 3 × 4, of the second one is 4 × 1, the size of product matrix will be 3 × 1, that is, the product matrix must have 3 rows and 1 column. According to the rule of multiplication of matrices, we get

$$\begin{pmatrix} 5 & 0 & 2 & 3 \\ 4 & 1 & 5 & 3 \\ 3 & 1 & -1 & 2 \end{pmatrix}\cdot\begin{pmatrix} 6 \\ -2 \\ 7 \\ 4 \end{pmatrix}=\begin{pmatrix} 5\cdot6+0\cdot(-2)+2\cdot7+3\cdot4 \\ 4\cdot6+1\cdot(-2)+5\cdot7+3\cdot4 \\ 3\cdot6+1\cdot(-2)+(-1)\cdot7+2\cdot4 \end{pmatrix}=\begin{pmatrix} 56 \\ 69 \\ 17 \end{pmatrix}.$$

Answer: $2)\ \begin{pmatrix} 1 & -2 & 16 & 6 \\ 11 & -7 & 29 & -9 \end{pmatrix};\ 3)\ (31);\ 4)\ \begin{pmatrix} -4 & 5 \\ -4 & 8 \end{pmatrix}.$

Problem 4. Knowing

$$A=\begin{pmatrix} 1 & 2 \\ -2 & -1 \\ 0 & 1 \end{pmatrix},\ B=\begin{pmatrix} 1 & -1 & 0 \\ 2 & 1 & -3 \end{pmatrix},\ C=\begin{pmatrix} 1 & 3 \\ 2 & 2 \\ 3 & 1 \end{pmatrix}.$$

find the matrices $AB–CB$ and $(A–C)\cdot B$ and compare them.

Answer: $AB-CB=(A-C)\cdot B=\begin{pmatrix} -2 & -1 & 3 \\ -10 & 1 & 9 \\ -3 & 3 & 0 \end{pmatrix}.$

Home tasks

Problem 5. Calculate:

$$1)\ \begin{pmatrix} 5 & 8 & -4 \\ 6 & 9 & -5 \\ 4 & 7 & -3 \end{pmatrix}\cdot\begin{pmatrix} 3 & 2 & 5 \\ 4 & -1 & 3 \\ 9 & 6 & 5 \end{pmatrix};\ 2)\ \begin{pmatrix} 3 & 2 & 1 \\ 0 & 1 & 2 \end{pmatrix}\cdot\begin{pmatrix} 1 \\ 2 \\ 3 \end{pmatrix};$$

$$3)\ \begin{pmatrix}2\\1\\3\end{pmatrix}\cdot\begin{pmatrix}1 & 2 & 3\end{pmatrix};\quad 4)\ \begin{pmatrix}-1 & 0\\1 & 2\end{pmatrix}^2.$$

Answer:

$$1)\ \begin{pmatrix}11 & -22 & 29\\9 & -27 & 32\\13 & -17 & 26\end{pmatrix};\quad 2)\ \begin{pmatrix}10\\8\end{pmatrix};\quad 3)\ \begin{pmatrix}2 & 4 & 6\\1 & 2 & 3\\3 & 6 & 9\end{pmatrix};\quad 4)\ \begin{pmatrix}1 & 0\\1 & 4\end{pmatrix}.$$

Problem 6. Calculate:

$$\begin{pmatrix}0 & 0 & 1\\1 & 1 & 2\\2 & 2 & 3\\3 & 3 & 4\end{pmatrix}\cdot\begin{pmatrix}-1 & -1\\2 & 2\\1 & 1\end{pmatrix}\cdot\begin{pmatrix}4\\1\end{pmatrix}.$$

Answer: $\begin{pmatrix}5\\15\\25\\35\end{pmatrix}$.

Problem 7. Knowing

$$A=\begin{pmatrix}1 & -1 & 2\\3 & -2 & -3\end{pmatrix},\quad B=\begin{pmatrix}0 & -2 & 1\\2 & -3 & -4\end{pmatrix},\quad C=\begin{pmatrix}1 & 0\\2 & 1\\3 & 2\end{pmatrix}.$$

find the matrix $AC{-}BC$.

Answer: $\begin{pmatrix}6 & 3\\6 & 3\end{pmatrix}$.

1.2 DETERMINANTS AND CALCULATION OF THEIR FEATURES

A certain number is associated with each square matrix and this number is said to be a determinant corresponding to this matrix. The number $a_{11}\,a_{22} - a_{12}\,a_{21}$ is a determinant corresponding to the second-order matrix

$$A = \begin{pmatrix} a_{11} & a_{12} \\ a_{21} & a_{22} \end{pmatrix}$$

and is denoted by

$$\det A = \begin{vmatrix} a_{11} & a_{12} \\ a_{21} & a_{22} \end{vmatrix} = a_{11}a_{22} - a_{12}a_{21}.$$

It is easy to remember the rule of calculation of the second-order determinant by the following scheme:

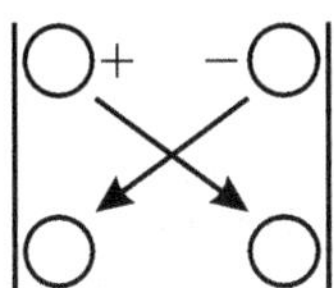

Here in front of the product of the elements along the principal diagonal we take the plus sign and in front of the product of elements along the auxiliary diagonal we take the minus sign

The number

$$a_{11}a_{22}a_{33} + a_{12}a_{23}a_{31} + a_{21}a_{32}a_{13} -$$
$$-a_{13}a_{22}a_{31} - a_{21}a_{12}a_{33} - a_{32}a_{23}a_{11}$$

is a determinant corresponding to the third-order matrix

$$A = \begin{pmatrix} a_{11} & a_{12} & a_{13} \\ a_{21} & a_{22} & a_{23} \\ a_{31} & a_{32} & a_{33} \end{pmatrix}$$

and is denoted by

$$\det A = \begin{vmatrix} a_{11} & a_{12} & a_{13} \\ a_{21} & a_{22} & a_{23} \\ a_{31} & a_{32} & a_{33} \end{vmatrix}.$$

It is expedient to remember the rule for calculation of the third-order determinant by the following scheme:

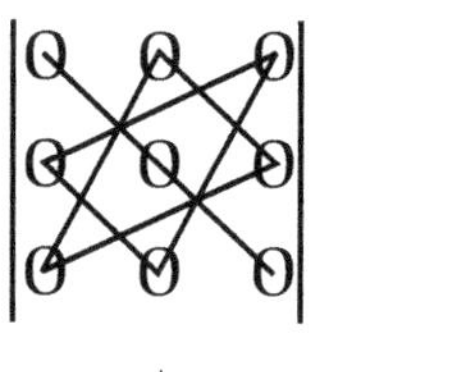 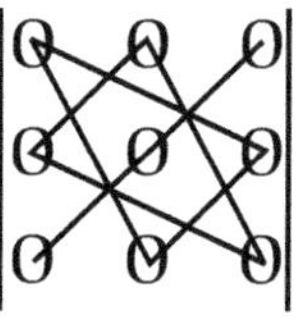

+ −

Now give definition of an arbitrary order determinant. For that we introduce some auxiliary denotation.

Any arrangement of first n number natural numbers 1, 2, 3, …., n is called substitution. The number of all possible substitutions made of n elements is $n! = 1 \cdot 2 \cdot \ldots \cdot n$ (n factorial). For example, substitutions made of the numbers 1, 2, 3 are

$$\begin{matrix} (1\,2\,3) & (2\,1\,3) & (3\,1\,2) \\ (1\,3\,2) & (2\,3\,1) & (3\,2\,1) \end{matrix}$$

Their amount is $3! = 1 \cdot 2 \cdot 3 = 6$.

In substitution, for $i > j$, if the number i precedes the number j -it is said that in this substitution the numbers i and j compose inversion. For example, in substitution (2 3 1), in spite of $2 > 1$, as 1 follows 2, the number 2 and 1 compose inversion. Just in the same way in this substitution the numbers 3 and 1 compose inversion.

The substitution with even number of total amount of inversions is called even substitution and substitution with odd number of total amount of inversions is called odd substitution. For example, as in substitution (5 3 1 2 6 4) the total amount of inversions is $1 + 2 + 2 + 2 = 7$, this is odd substitution. As in substitution (4 3 1 2 6 5) the amount of inversions is $1 + 2 + 2 + 1 = 6$, this is even substitution.

Taking only one element from every row and column of n-th order square matrix

$$A = \begin{pmatrix} a_{11} & a_{12} & . & . & . & a_{1n} \\ a_{21} & a_{22} & . & . & . & a_{2n} \\ . & . & . & . & . & . \\ a_{n1} & a_{n2} & . & . & . & a_{nn} \end{pmatrix},$$

we make the following product:

$$a_{1\alpha_1} a_{2\alpha_2} \cdots a_{n\alpha_n} \cdot \tag{1.1}$$

Here, α_1 denotes the number of column of the element taken from the first row, α_2 denotes the number of the column of the element taken from the second row, etc. α_n denotes the number of column of the element taken from the n-th row. It is clear that the number of all possible products in the form (1.1) equals the number of substitutions composed of the numbers 1, 2, 3,... and composing the second indices in (1.1). The number of such substitutions is n. When the substitution $(\alpha_1\ \alpha_2\ \ldots\ \alpha_n)$ is even, in front of the term (1.1) we will take the plus sign, when it is odd we will take the minus sign.

Definition. Taking only one element from every row and column of the matrix A and putting the appropriate sign in front of them, the sum of $n!$ number terms in the form (1.1) is said to be n-th order determinant corresponding to the matrix A and is denoted by

$$\det A = \begin{vmatrix} a_{11} & a_{12} & . & . & . & a_{1n} \\ a_{21} & a_{22} & . & . & . & a_{2n} \\ . & . & . & . & . & . \\ a_{n1} & a_{n2} & . & . & . & a_{nn} \end{vmatrix}.$$

Determinant has the following features.

Feature 1. If we permutate the rows and appropriate columns of the determinant, the value of the determinant does not change.

Feature 2. If we permutate two rows and two columns of the determinant, it changes only its sign.

Feature 3. The determinant with same two rows or columns equals zero.

Feature 4. Common factor of all elements of any row or column of determinant may be taken out of the sign of determinant.

Feature 5. If a determinant has a row or a column whose all elements are zero, this determinant equals zero.

Feature 6. If a determinant has proportional rows or columns, this determinant equals zero.

Feature 7. If a determinant has such a row or column that all its elements are in the form of the sum of two numbers, then this determinant equals the sum of such two determinants that the first addends are written in this row and column of the first determinant, and the second addends in the same row and column of the second determinant, the remaining rows and columns of both determinants are identical with appropriate rows and columns of the given determinant.

Feature 8. Having multiplied all elements of any row and column of a determinant by a certain number and adding to appropriate elements of another row and column, the value of the determinant does not change.

In the n-th order determinant, the $(n-1)$-th order determinant obtained by rubbing out the i-th row and the j-th column, where the element a_{ij} stands, is said to be a minor of the element a_{ij} and is denoted by M_{ij}.

The number A_{ij} determined by the formula $A_{ij} = (-1)^{i+j} M_{ij}$ is called a cofactor of the element a_{ij}. For example, in third-order determinant

$$\begin{vmatrix} 1 & 2 & -3 \\ 5 & 4 & 6 \\ -1 & -2 & -4 \end{vmatrix}$$

the minor of the number −2 is the second-order determinant

$$M_{32} = \begin{vmatrix} 1 & -3 \\ 5 & 6 \end{vmatrix} = 6 + 15 = 21,$$

obtained by rubbing out the third row and the second column, its cofactor is the number

$$A_{32} = (-1)^{3+2} M_{32} = (-1)^5 \cdot 21 = -21.$$

The following statement is true.

Theorem. The sum of products of all elements of any row or column by their own cofactor equals this determinant.

This theorem may be expressed by this formula:

$$\det A = a_{i1} A_{i1} + a_{i2} A_{i2} + \ldots + a_{in} A_{in}; \tag{1.2}$$

$$\det A = a_{1j} A_{1j} + a_{2j} A_{2j} + \ldots + a_{nj} A_{nj}. \tag{1.3}$$

Formula (1.2) is said to be expansion formula with respect to the i-th row elements of the determinant, formula (1.3) is said to be expansion formula with respect to the j-th column elements.

Problems to be solved in auditorium

Problem 8. Calculate:

$$1)\ \begin{vmatrix} 1 & 2 \\ 3 & 4 \end{vmatrix};\quad 2)\ \begin{vmatrix} 2 & -1 \\ 3 & 4 \end{vmatrix};\quad 3)\ \begin{vmatrix} -1 & 4 \\ -2 & -3 \end{vmatrix};\quad 4)\ \begin{vmatrix} \sin\alpha & \cos\alpha \\ -\cos\alpha & \sin\alpha \end{vmatrix}.$$

Answer: 1) −2; 2) 11; 3) 11; 4) 1.

Problem 9. Solve the equation:

$$\begin{vmatrix} \cos 8x & -\sin 5x \\ \sin 8x & \cos 5x \end{vmatrix} = 0.$$

Answer: $\frac{\pi}{6}+\frac{k\pi}{3},\ k \in Z.$

Problem 10. Calculate:

$$1)\ \begin{vmatrix} 1 & 1 & 1 \\ 1 & 2 & 3 \\ 1 & 3 & 6 \end{vmatrix};\quad 2)\ \begin{vmatrix} 3 & 4 & -5 \\ 8 & 7 & -2 \\ 2 & -1 & 8 \end{vmatrix};\quad 3)\ \begin{vmatrix} 1 & -1 & 1 \\ 1 & 2 & 3 \\ 0 & 1 & 2 \end{vmatrix}.$$

Answer: 1) 1; 2) 0; 3) 4.

Problem 11. Not opening the determinant proves the validity of the identity:

$$\begin{vmatrix} a_1+b_1x & a_1-b_1x & c_1 \\ a_2+b_2x & a_2-b_2x & c_2 \\ a_3+b_3x & a_3-b_3x & c_3 \end{vmatrix} = -2x \begin{vmatrix} a_1 & b_1 & c_1 \\ a_2 & b_2 & c_2 \\ a_3 & b_3 & c_3 \end{vmatrix}.$$

***Guideline*:** Add the second column of the determinant in the left side to the first column, take the second one from the first column of the obtained determinant out of the sign of determinant, then multiply this column by −1, add to the second column and take the x outside of the sign of determinant.

Problem 12. Not opening, using the features of a determinant, calculate the following determinant:

$$\begin{vmatrix} x+y & z & 1 \\ y+z & x & 1 \\ z+x & y & 1 \end{vmatrix}.$$

Answer: 0.

Problem 13. Formulate the following determinants in the convenient form and calculate them separating in row and column elements:

$$1)\ \begin{vmatrix} 1 & 2 & 5 \\ 3 & -4 & 7 \\ -3 & 12 & -15 \end{vmatrix}; \quad 2)\ \begin{vmatrix} 1 & 1 & 1 \\ \omega_1 & \omega_2 & \omega_3 \\ \omega_1^2 & \omega_2^2 & \omega_3^2 \end{vmatrix}$$

Solution of 2): Multiply the third column by -1 and add it to the first and second column. Take $(\omega_1 - \omega_3)$ from the first column of the determinant and $(\omega_2 - \omega_3)$ from the second column out of the sign of determinant as a common factor:

$$\begin{vmatrix} 1 & 1 & 1 \\ \omega_1 & \omega_2 & \omega_3 \\ \omega_1^2 & \omega_2^2 & \omega_3^2 \end{vmatrix} = \begin{vmatrix} 0 & 0 & 1 \\ \omega_1 - \omega_3 & \omega_2 - \omega_3 & \omega_3 \\ \omega_1^2 - \omega_3^2 & \omega_2^2 - \omega_3^2 & \omega_3^2 \end{vmatrix} =$$

$$= (\omega_1 - \omega_3)(\omega_2 - \omega_3) \begin{vmatrix} 0 & 0 & 1 \\ 1 & 1 & \omega_3 \\ \omega_1 + \omega_3 & \omega_2 + \omega_3 & \omega_3^2 \end{vmatrix} =$$

$$= (\omega_1 - \omega_3)(\omega_2 - \omega_3)(-1)^{1+3} \cdot 1 \cdot \begin{vmatrix} 1 & 1 \\ \omega_1 + \omega_3 & \omega_2 + \omega_3 \end{vmatrix} =$$

$$= (\omega_1 - \omega_3)(\omega_2 - \omega_3)(\omega_2 + \omega_3 - \omega_1 - \omega_3) =$$

$$= (\omega_1 - \omega_3)(\omega_2 - \omega_3)(\omega_2 - \omega_1) =$$

$$= -(\omega_1 - \omega_2)(\omega_1 - \omega_3)(\omega_2 - \omega_3).$$

Answer: 1) 144. ***Guideline*:** You can multiply the first row by -3 and add to the second row, multiply by 3, and add to the third row.

Problem 14. Calculate the following determinant:

$$1)\begin{vmatrix}1&0&0&0\\3&2&-1&7\\0&0&3&5\\0&0&0&4\end{vmatrix};\ 2)\begin{vmatrix}2&3&-3&4\\2&1&-1&2\\6&2&1&0\\2&3&0&-5\end{vmatrix};\ 3)\begin{vmatrix}3&-1&4&2\\5&2&0&1\\0&2&1&-3\\6&-2&9&8\end{vmatrix}.$$

Answer: 1) 24; 2) 48; 3) 223.

Home tasks

Problem 15. Calculate:

$$1)\begin{vmatrix}5&6\\3&4\end{vmatrix};\quad 2)\begin{vmatrix}3&2\\-2&-1\end{vmatrix};\quad 3)\begin{vmatrix}\cos\alpha&\sin\alpha\\\sin\alpha&\cos\alpha\end{vmatrix}.$$

Answer: 1) 2; 2) 1; 3) $\cos 2\alpha$.

Problem 16. Solve the equation:

$$\begin{vmatrix}x&x+1\\-4&x+1\end{vmatrix}=0.$$

Answer: $x_1=-4, x_2=-1.$

Problem 17. Calculate:

$$1)\begin{vmatrix}1&2&3\\4&5&6\\7&8&9\end{vmatrix};\quad 2)\begin{vmatrix}1&1&1\\-1&0&1\\-1&-1&0\end{vmatrix};\quad 3)\begin{vmatrix}a&b&c\\1&-1&2\\1&2&3\end{vmatrix}.$$

Answer: 1) 0; 2) 1; 3) $-7a-\mathrm{b}+3\mathrm{c}$.

Problem 18. Not opening the determinant proves the identity:

$$\begin{vmatrix}a_1+b_1x&a_1x+b_1&c_1\\a_2+b_2x&a_2x+b_2&c_2\\a_3+b_3x&a_3x+b_3&c_3\end{vmatrix}=(1-x^2)\begin{vmatrix}a_1&b_1&c_1\\a_2&b_2&c_2\\a_3&b_3&c_3\end{vmatrix}.$$

Guideline: From the second column of the determinant obtained by multiplying the first row of the determinant in the left-hand side by $-x$, put the

$(1 - x^2)$ out the sign of determinant. Then multiply the second column of the obtained last determinant by $-x$ and add to the first column.

Problem 19. Calculate:

$$1)\begin{vmatrix} -2 & 0 & 3 & 5 \\ 7 & 1 & 9 & 11 \\ 0 & 0 & 3 & 13 \\ 0 & 0 & 0 & 5 \end{vmatrix}; \quad 2)\begin{vmatrix} 2 & -1 & 1 & 0 \\ 0 & 1 & 2 & -1 \\ 3 & -1 & 2 & 3 \\ 3 & 1 & 6 & 1 \end{vmatrix}; \quad 3)\begin{vmatrix} 1 & 1 & 1 & 1 \\ 1 & 2 & 3 & 4 \\ 1 & 4 & 9 & 16 \\ 1 & 8 & 27 & 64 \end{vmatrix}.$$

Answer: 1) –30; 2) 0; 3) 12.

1.3 RANK OF MATRICES AND ITS CALCULATION RULES

The k-th order determinant made of the elements standing in the intersection of k number rows and k number columns satisfying the condition $k \leq \min\{m, n\}$ in $m \times n$ dimensional matrix A is called k-th order minor of the matrix A. For example, in the 3×4 dimensional matrix

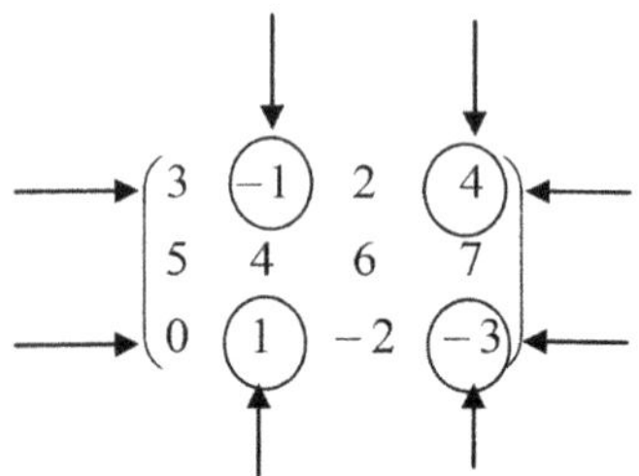

if take two rows (e.g., first and third) and two columns (e.g., second and fourth), the second-order determinant

$$\begin{vmatrix} -1 & 4 \\ 1 & -3 \end{vmatrix} = 3 - 4 = -1$$

made of elements standing at their intersection, will be a second-order minor of the given matrix.

Definition. Order of the highest order non-zero minors of the matrix is said to be the rank of this matrix.

The rank of the matrix A is denoted as rank $A = r$. If the rank of the matrix equals r, this means that at least one of the r-th order minors of this matrix is non-zero, all minors with orders higher than r equal zero.

For calculating the rank of a matrix, two methods are used. By the first method, the highest order non-zero minor of a matrix is found. This time it should be taken into account that if all k-th order minors equal zero, then all $(k + 1)$-th order minors will be equal to zero as well.

In the second method for calculation of the rank of a matrix, by means of elementary transformations it is shaped so that the rank of the obtained matrix is directly determined. The following operations carried out on rows and columns of a matrix are called elementary transformations of a matrix:

1. Permutation of any rows and columns of a matrix;
2. Multiplication or division of all elements of any row or column of a matrix to the same nonzero number;
3. Multiplication of all elements of any row or column of a matrix by the same number and addition to appropriate element of other row or column elements.

The following theorem is true.

Theorem 1. Elementary transformations do not change the rank of a matrix.

If except the elements $a_{11}, a_{22}, \ldots, a_{rr}$ $(1 \leq r \min \{m, n\})$ standing in the principal diagonal of $m \times n$ dimensional matrix A all other elements equal zero, then A is called a diagonal matrix. For example,

$$\begin{pmatrix} 2 & 0 & 0 & 0 \\ 0 & 1 & 0 & 0 \\ 0 & 0 & 0 & 0 \end{pmatrix}$$

is a diagonal matrix.

Theorem 2. By elementary transformations, any matrix may be reduced to a diagonal form.

It is clear that the rank of any matrix in diagonal form equals the number of nonzero elements in the principal diagonal.

The step matrix is such a matrix that beginning from the second row, in each row the column number of nonzero two elements is greater than the column number of nonzero two elements in the preceding row. For example, the matrix

$$A=\begin{pmatrix} 0 & 1 & 3 & 4 & -7 \\ 0 & 0 & 0 & 5 & 6 \\ 0 & 0 & 0 & 0 & 3 \\ 0 & 0 & 0 & 0 & 0 \end{pmatrix}$$

is in the step form. By elementary transformations it is easy to reduce a step-matrix to a diagonal form. This time, the number of non-zero elements in the principal diagonal of the matrix will be equal to the number of nonzero rows of the step-matrix. Therefore, the rank of a step-matrix equals the number of its nonzero rows.

If we multiply the second column of the above-given matrix by −3 and add to the third column, and move the first and third columns to the last columns by permutation, we get:

$$\begin{pmatrix} 0 & 1 & 3 & 4 & -7 \\ 0 & 0 & 0 & 5 & 6 \\ 0 & 0 & 0 & 0 & 3 \\ 0 & 0 & 0 & 0 & 0 \end{pmatrix} \to \begin{pmatrix} 0 & 1 & 0 & 4 & -7 \\ 0 & 0 & 0 & 5 & 6 \\ 0 & 0 & 0 & 0 & 3 \\ 0 & 0 & 0 & 0 & 0 \end{pmatrix} \to \begin{pmatrix} 1 & 4 & -7 & 0 & 0 \\ 0 & 5 & 6 & 0 & 0 \\ 0 & 0 & 3 & 0 & 0 \\ 0 & 0 & 0 & 0 & 0 \end{pmatrix}.$$

In the obtained last matrix, all elements standing in one side of the principal diagonal are equal to zero. Such matrices are called triangular matrices and their rank is equal to the number of nonzero elements in the principal diagonal (rank $A = 3$).

Problems to be solved in auditorium

Problem 20. Find the rank of the matrix:

$$1)\ A=\begin{pmatrix} 1 & -1 & 2 & 3 \\ 2 & -2 & 4 & 6 \\ -3 & 3 & -6 & -9 \end{pmatrix};\quad 2)\ B=\begin{pmatrix} -1 & 4 & 1 & -2 \\ -2 & 8 & 2 & -4 \\ -3 & 12 & 3 & -6 \end{pmatrix}.$$

***Guideline*:** As the rows of the matrix A, the columns of B are proportional; all second and third-order minors are equal to zero.

Answer: 1) rank $A = 1$; 2) rank $B = 1$.

Problem 21. By choosing a non-zero, highest order minor, find the rank of the matrix:

$$1)\begin{pmatrix}1&2\\3&6\end{pmatrix};\quad 2)\begin{pmatrix}2&4\\5&3\end{pmatrix};\quad 3)\begin{pmatrix}-1&2&3\\1&-2&5\\0&4&-4\end{pmatrix};$$

$$4)\begin{pmatrix}1&2&3\\1&1&1\\-1&0&1\end{pmatrix};\quad 5)\begin{pmatrix}4&3&2&2\\0&2&1&1\\0&0&3&3\end{pmatrix};\quad 6)\begin{pmatrix}-1&1&2&3&4\\-2&0&2&3&5\\-3&0&0&2&6\end{pmatrix}.$$

Answer: 1) 1; 2) 2; 3) 3; 4) 2; 5) 3; 6) 3.

Problem 22. Find the rank of the matrix by elementary transformations:

$$1)\begin{pmatrix}1&2&3\\3&7&10\\3&7&11\end{pmatrix};\qquad 2)\begin{pmatrix}0&0&1&-1\\0&3&1&4\\2&7&6&-1\\1&2&2&-1\end{pmatrix};$$

$$3)\begin{pmatrix}5&3&4&1&2\\10&7&7&2&9\\10&6&8&3&6\\10&9&12&5&10\end{pmatrix};\qquad 4)\begin{pmatrix}0&3&4&5&6\\0&1&-8&-5&-2\\0&1&6&5&4\\0&0&2&-3&-5\end{pmatrix}.$$

Solution of 4): In the given matrix, permutate at first the first row and second row, then in the obtained matrix permutate the second and the fourth row, and reduce the matrix to the step form:

$$\begin{pmatrix}0&3&4&5&6\\0&1&-8&-5&-2\\0&1&6&5&4\\0&0&2&-3&-5\end{pmatrix}\to\begin{pmatrix}0&1&-8&-5&-2\\0&0&2&-3&-5\\0&1&6&5&4\\0&3&4&5&6\end{pmatrix}\to$$

$$\to\begin{pmatrix}0&1&-8&-5&-2\\0&0&2&-3&-5\\0&0&14&10&6\\0&0&28&20&12\end{pmatrix}\to\begin{pmatrix}0&1&-8&-5&-2\\0&0&2&-3&-5\\0&0&0&31&41\\0&0&0&62&82\end{pmatrix}\to\begin{pmatrix}0&1&-8&-5&-2\\0&0&2&-3&-5\\0&0&0&31&41\\0&0&0&0&0\end{pmatrix}.$$

In the last step-matrix, as the number of all rows with non-zero elements is three, the rank of the given matrix is 3.

Answer: 1) 3; 2) 4; 3) 4.

Home tasks

Problem 23. Find the rank of the matrix by choosing a non-zero highest order minor:

$$1)\ \begin{pmatrix} 2 & 3 \\ 6 & 9 \end{pmatrix};\quad 2)\ \begin{pmatrix} 1 & -1 \\ 4 & 3 \end{pmatrix};\quad 3)\ \begin{pmatrix} 1 & 2 & 3 \\ 1 & -1 & -4 \\ 2 & 1 & -1 \end{pmatrix};\quad 4)\ \begin{pmatrix} 2 & -4 & 3 \\ 1 & -2 & 1 \\ 0 & 1 & -1 \end{pmatrix};$$

$$5)\ \begin{pmatrix} 2 & -1 & 3 & -2 & 4 \\ 4 & -2 & 5 & 1 & 7 \\ 2 & -1 & 1 & 8 & 2 \end{pmatrix};\quad 6)\ \begin{pmatrix} 3 & -1 & 3 & 2 & 5 \\ 5 & -3 & 2 & 3 & 4 \\ 1 & -3 & -5 & 0 & -7 \\ 7 & -5 & 1 & 4 & 1 \end{pmatrix}.$$

Answer: 1) 1; 2) 2; 3) 2; 4) 3; 5) 2; 6) 3.

Problem 24. Find the rank of the matrix by elementary transformations:

$$1)\ \begin{pmatrix} 25 & 31 & 17 & 43 \\ 75 & 94 & 53 & 132 \\ 75 & 94 & 54 & 134 \\ 25 & 32 & 20 & 48 \end{pmatrix};\quad 2)\ \begin{pmatrix} -1 & 3 & 3 & -4 \\ 4 & -7 & -2 & 1 \\ -3 & 5 & 1 & 0 \\ -2 & 3 & 0 & 1 \end{pmatrix};$$

$$3)\ \begin{pmatrix} 0 & 1 & 1 & 0 & 0 \\ 1 & 1 & 0 & 0 & 0 \\ 0 & 1 & 0 & 1 & 1 \\ 1 & 0 & 1 & 0 & 0 \\ 0 & 0 & 1 & 1 & 0 \end{pmatrix};\quad 4)\ \begin{pmatrix} 0 & 1 & 0 & 4 & 3 & 1 \\ 0 & 1 & 3 & 0 & 2 & 1 \\ 2 & 1 & 0 & 0 & 1 & 1 \\ -1 & 2 & -1 & -1 & -1 & 1 \end{pmatrix}.$$

Answer: 1) 3; 2) 2; 3) 5; 4) 4.

1.4 INVERSE MATRIX AND METHODS FOR ITS FINDING

Definition. If there exists a matrix X for n-th order square matrix A such that

$$AX = XA = E, \tag{1.4}$$

X is said to be the inverse matrix of A and is denoted by A^{-1}. Here E denotes the n-th order unique square matrix.

It is clear that if there exists a matrix X satisfying (1.4), then it also will be an n-th order square matrix.

Theorem. If the determinant of n-th order square matrix A is non-zero, then A has an inverse matrix and this inverse matrix is determined by the following formula:

$$A^{-1} = \frac{1}{\det A}\begin{pmatrix} A_{11} & A_{21} & \cdot & \cdot & \cdot & A_{n1} \\ A_{12} & A_{22} & \cdot & \cdot & \cdot & A_{n2} \\ \cdot & \cdot & \cdot & \cdot & \cdot & \cdot \\ A_{1n} & A_{2n} & \cdot & \cdot & \cdot & A_{nn} \end{pmatrix}. \tag{1.5}$$

Here, A_{ij} denotes the cofactor of a_{ij} elements of the determinant corresponding to the matrix A.

According to this theorem, the inverse matrix of the matrix A (if exists) must be found by the following sequence:

1. $\det A$ is calculated. If $\det A = 0$, then A has no inverse matrix;
2. Cofactor A_{ij} of each a_{ij} element of $\det A$ is calculated;
3. In A instead of each a_{ij} element, its cofactor A_{ij} is written;
4. The rows of the matrix obtained in the previous step are transposed with appropriate columns, that is, the matrix is transposed;
5. All terms of the matrix obtained in the fourth step are divided by $\det A$.

Note that instead of fourth and first steps, after calculating the cofactors, we can directly use formula (1.5) and find A^{-1}.

As is seen, in order to calculate the inverse matrix of n-th order matrix it is necessary to calculate one n-th order ($\det A$) and n^2 the number of $(n-1)$-th order (A_{ij}) determinants. This takes much time and calculations. Therefore, in order to calculate an inverse matrix the following matrix based on only elementary transformations of a matrix is much used.

For finding the inverse matrix of n-th order matrix A we write the n-th order unique matrix E in its right-hand side and obtain $n \times (2n)$-dimensional rectangular matrix. By making transformations only on rows of these rectangular matrices one can succeed to get a unique matrix E in place of A. This time, the new matrix obtained in place of E is an inverse matrix of the matrix A. We can new matrix write the above operations as a scheme in the following way

$(A|E)\to(E|A^{-1})$. Find the inverse matrix of the matrix $A=\begin{pmatrix}1 & 2\\ 3 & 4\end{pmatrix}$ by this method. As det$A = 4 - 6 = -2 \neq 0$, an inverse matrix exists.

$$(A|E)=\left(\begin{array}{cc|cc}1 & 2 & 1 & 0\\ 3 & 4 & 0 & 1\end{array}\right)\to$$

(multiply the first row by -3 and add it to the second row)

$$\to\left(\begin{array}{cc|cc}1 & 2 & 1 & 0\\ 0 & -2 & -3 & 1\end{array}\right)\to$$

(add the second row to the first row)

$$\to\left(\begin{array}{cc|cc}1 & 0 & -2 & 1\\ 0 & -2 & -3 & 1\end{array}\right)\to$$

(divide all elements of the second row into -2)

$$\to\left(\begin{array}{cc|cc}1 & 0 & -2 & 1\\ 0 & 1 & \dfrac{3}{2} & -\dfrac{1}{2}\end{array}\right)\Rightarrow A^{-1}=\begin{pmatrix}-2 & 1\\ \dfrac{3}{2} & -\dfrac{1}{2}\end{pmatrix}.$$

Problems to be solved in auditorium

Problem 25. By the method of calculation of cofactors, find the inverse matrix of the matrix:

$$1)\begin{pmatrix}1 & 2\\ 3 & 4\end{pmatrix};\quad 2)\begin{pmatrix}3 & 4\\ 5 & 7\end{pmatrix};\quad 3)\begin{pmatrix}2 & 5 & 7\\ 6 & 3 & 4\\ 5 & -2 & -3\end{pmatrix};\quad 4)\begin{pmatrix}3 & -4 & 5\\ 2 & -3 & 1\\ 3 & -5 & -1\end{pmatrix}.$$

Solution of 4): At first, we calculate the determinant appropriate to the matrix:

$$\begin{vmatrix} 3 & -4 & 5 \\ 2 & -3 & 1 \\ 3 & -5 & -1 \end{vmatrix} = 9 - 12 - 50 + 45 - 8 + 15 = -1.$$

Find the cofactors of the elements of this determinant:

$$A_{11} = (-1)^{1+1} \cdot \begin{vmatrix} -3 & 1 \\ 5 & -1 \end{vmatrix} = 3 + 5 = 8; \quad A_{12} = (-1)^{1+2} \cdot \begin{vmatrix} 2 & 1 \\ 3 & -1 \end{vmatrix} = -(-2-3) = 5;$$

$$A_{13} = (-1)^{1+3} \cdot \begin{vmatrix} 2 & -3 \\ 3 & -5 \end{vmatrix} = -10 + 9 = -1; \; A_{21} = (-1)^{2+1} \cdot \begin{vmatrix} -4 & 5 \\ -5 & -1 \end{vmatrix} = -29;$$

$$A_{22} = (-1)^{2+2} \cdot \begin{vmatrix} 3 & 5 \\ 3 & -1 \end{vmatrix} = -3 - 15 = -18; \; A_{23} = (-1)^{2+3} \cdot \begin{vmatrix} 3 & -4 \\ 3 & -5 \end{vmatrix} = 3;$$

$$A_{31} = (-1)^{3+1} \cdot \begin{vmatrix} -4 & 5 \\ -3 & 1 \end{vmatrix} = -4 + 15 = 11; \quad A_{32} = (-1)^{3+2} \cdot \begin{vmatrix} 3 & 5 \\ 2 & 1 \end{vmatrix} = -(3-10) = 7;$$

$$A_{33} = (-1)^{3+3} \cdot \begin{vmatrix} 3 & -4 \\ 2 & -3 \end{vmatrix} = -9 + 8 = -1.$$

Now find the inverse matrix by formula (1.5):

$$A^{-1} = \frac{1}{\det A} \cdot \begin{pmatrix} A_{11} & A_{21} & A_{31} \\ A_{12} & A_{22} & A_{32} \\ A_{13} & A_{23} & A_{33} \end{pmatrix} = \frac{1}{-1} \cdot \begin{pmatrix} 8 & -29 & 11 \\ 5 & -18 & 7 \\ -1 & 3 & -1 \end{pmatrix} = \begin{pmatrix} -8 & 29 & -11 \\ -5 & 18 & -7 \\ 1 & -3 & 1 \end{pmatrix}.$$

Answer:

$$1) \begin{pmatrix} -2 & 1 \\ \frac{3}{2} & -\frac{1}{2} \end{pmatrix}; \; 2) \begin{pmatrix} 7 & -4 \\ -5 & 3 \end{pmatrix}; \; 3) \begin{pmatrix} 1 & -1 & 1 \\ -38 & 41 & -34 \\ 27 & -29 & 24 \end{pmatrix}.$$

Problem 26. Find inverse matrices of the following matrix by elementary transformations

$$1) \; A = \begin{pmatrix} 2 & 7 & 3 \\ 3 & 9 & 4 \\ 1 & 5 & 3 \end{pmatrix}; \; 2) \; B = \begin{pmatrix} 1 & 2 & 3 \\ 2 & 3 & 5 \\ 2 & 5 & 8 \end{pmatrix}; \; 3) \; C = \begin{pmatrix} 1 & 1 & 1 & 1 \\ 1 & 1 & -1 & -1 \\ 1 & -1 & 1 & -1 \\ 1 & -1 & -1 & 1 \end{pmatrix}.$$

Solution of 1):

$$\left(\begin{array}{ccc|ccc} 2 & 7 & 3 & 1 & 0 & 0 \\ 3 & 9 & 4 & 0 & 1 & 0 \\ 1 & 5 & 3 & 0 & 0 & 1 \end{array}\right) \to \left(\begin{array}{ccc|ccc} 1 & 5 & 3 & 0 & 0 & 1 \\ 2 & 7 & 3 & 1 & 0 & 0 \\ 3 & 9 & 4 & 0 & 1 & 0 \end{array}\right) \to$$

$$\to \left(\begin{array}{ccc|ccc} 1 & 5 & 3 & 0 & 0 & 1 \\ 0 & -3 & -3 & 1 & 0 & -2 \\ 0 & -6 & -5 & 0 & 1 & -3 \end{array}\right) \to \left(\begin{array}{ccc|ccc} 1 & 5 & 3 & 0 & 0 & 1 \\ 0 & -3 & -3 & 1 & 0 & -2 \\ 0 & 0 & 1 & -2 & 1 & 1 \end{array}\right) \to$$

$$\to \left(\begin{array}{ccc|ccc} 1 & 5 & 0 & 6 & -3 & -2 \\ 0 & -3 & 0 & -5 & 3 & 1 \\ 0 & 0 & 1 & -2 & 1 & 1 \end{array}\right) \to \left(\begin{array}{ccc|ccc} 1 & 5 & 0 & 6 & -3 & -2 \\ 0 & 1 & 0 & \frac{5}{3} & -1 & -\frac{1}{3} \\ 0 & 0 & 1 & -2 & 1 & 1 \end{array}\right) \to$$

$$\to \left(\begin{array}{ccc|ccc} 1 & 0 & 0 & -\frac{7}{3} & 2 & -\frac{1}{3} \\ 0 & 1 & 0 & \frac{5}{3} & -1 & -\frac{1}{3} \\ 0 & 0 & 1 & -2 & 1 & 1 \end{array}\right) \Rightarrow {}^{-1} = \begin{pmatrix} -\frac{7}{3} & 2 & -\frac{1}{3} \\ \frac{5}{3} & -1 & -\frac{1}{3} \\ -2 & 1 & 1 \end{pmatrix}.$$

Answer: 2) $\begin{pmatrix} 1 & 1 & -1 \\ 6 & 2 & -1 \\ -4 & 1 & 1 \end{pmatrix}$; 3) $\frac{1}{4} \cdot \begin{pmatrix} 1 & 1 & 1 & 1 \\ 1 & 1 & -1 & -1 \\ 1 & -1 & 1 & -1 \\ 1 & -1 & -1 & 1 \end{pmatrix}$.

Problem 27. Solve the matrix equations:

$$1)\ X \cdot \begin{pmatrix} 3 & -2 \\ 5 & -4 \end{pmatrix} = \begin{pmatrix} -1 & 2 \\ -5 & 6 \end{pmatrix}; \quad 2)\ Y \cdot \begin{pmatrix} 5 & 3 & 1 \\ 1 & -3 & -2 \\ -5 & 2 & 1 \end{pmatrix} = \begin{pmatrix} -8 & 3 & 0 \\ -5 & 9 & 0 \\ -2 & 15 & 0 \end{pmatrix}.$$

Answer: 1) $X = \begin{pmatrix} 3 & -2 \\ 5 & -4 \end{pmatrix}$; 2) $Y = \begin{pmatrix} 1 & 2 & 3 \\ 4 & 5 & 6 \\ 7 & 8 & 9 \end{pmatrix}$.

Home tasks

Problem 28. Find inverse matrices of the given matrices:

$$1)\begin{pmatrix} -1 & 1 \\ 2 & 1 \end{pmatrix}; 2)\begin{pmatrix} 3 & -2 \\ 1 & -1 \end{pmatrix}; 3)\begin{pmatrix} 2 & 5 & 7 \\ 6 & 3 & 4 \\ 5 & -2 & -3 \end{pmatrix}; 4)\begin{pmatrix} 0 & 0 & 1 & -1 \\ 0 & 3 & 1 & 4 \\ 2 & 7 & 6 & -1 \\ 1 & 2 & 2 & -1 \end{pmatrix}.$$

Answer:

$$1)\begin{pmatrix} -\frac{1}{3} & \frac{1}{3} \\ \frac{2}{3} & \frac{1}{3} \end{pmatrix}; 2)\begin{pmatrix} 1 & -2 \\ 1 & -3 \end{pmatrix}; 3)\begin{pmatrix} 1 & -1 & 1 \\ -38 & 41 & -34 \\ 27 & -29 & 24 \end{pmatrix}; 4)\begin{pmatrix} -\frac{1}{6} & \frac{1}{2} & -\frac{7}{6} & \frac{10}{3} \\ -\frac{7}{6} & -\frac{1}{2} & \frac{5}{6} & -\frac{5}{3} \\ \frac{3}{2} & \frac{1}{2} & -\frac{1}{2} & 1 \\ \frac{1}{2} & \frac{1}{2} & -\frac{1}{2} & 1 \end{pmatrix}.$$

Problem 29. Solve the matrix equations:

$$1)\begin{pmatrix} 1 & 2 \\ 3 & 4 \end{pmatrix} \cdot \% = \begin{pmatrix} 3 & 5 \\ 5 & 9 \end{pmatrix}; 2)\begin{pmatrix} 1 & 2 & -3 \\ 3 & 2 & -4 \\ 2 & -1 & 0 \end{pmatrix} \cdot \# = \begin{pmatrix} 1 & -3 & 0 \\ 10 & 2 & 7 \\ 10 & 7 & 8 \end{pmatrix}.$$

Answer: 1) $X = \begin{pmatrix} -1 & -1 \\ 2 & 3 \end{pmatrix}$; 2) $Y = \begin{pmatrix} 6 & 4 & 5 \\ 2 & 1 & 2 \\ 3 & 3 & 3 \end{pmatrix}$.

1.5 SYSTEM OF LINEAR EQUATIONS

A system of linear equations consisting of m number equations with n number unknowns may be written as follows:

$$\begin{cases} a_{11}x_1 + a_{12}x_2 + \ldots + a_{1n}x_n = b_1, \\ a_{21}x_1 + a_{22}x_2 + \ldots + a_{2n}x_n = b_2, \\ \cdot\ \cdot\ \cdot\ \cdot\ \cdot\ \cdot\ \cdot\ \cdot\ \cdot\ \cdot\ \cdot\ \cdot\ \cdot \\ a_{m1}x_1 + a_{m2}x_2 + \ldots + a_{mn}x_n = b_m. \end{cases} \tag{1.6}$$

A system with at least one solution is called a compatible or joint system. Make the following two matrices from a_{ij} coefficients of eq 1.6 and free members $b_1, b_2, \ldots, b_m$:

$$A = \begin{pmatrix} a_{11} & a_{12} & . & . & . & a_{1n} \\ a_{21} & a_{22} & . & . & . & a_{2n} \\ . & . & . & . & . & . \\ a_{m1} & a_{m2} & . & . & . & a_{mn} \end{pmatrix}, \qquad A' = \begin{pmatrix} a_{11} & a_{12} & . & . & . & a_{1n} & b_1 \\ a_{21} & a_{22} & . & . & . & a_{2n} & b_2 \\ . & . & . & . & . & . & . \\ a_{m1} & a_{m2} & . & . & . & a_{mn} & b_m \end{pmatrix}$$

A is called the main, A' the augmented matrix.

Theorem *(of Kronecker–Capelli).* Necessary and sufficient condition for this system of linear equations to be joint is equality of the rank of the main matrix of this system to the rank of the augmented matrix.

The following operations made on eq. (1.1) are called elementary transformations of the system of linear equations:

1. displacement of any two equations of the system;
2. multiplication or division of all terms standing in both hand side of any equation of the system by the same number;
3. multiplication of any of the equations of the system by a certain number and addition by other equation.

Theorem. The new system obtained by elementary transformations on the system of linear equations is equivalent to the previous system.

It is clear that elementary transformations made on the system of linear equations may be replaced by appropriate transformations on the rows of the augmented matrix of this system. One of the more general methods for solving the system of linear equations is the Gauss method. The essence of this method is that by elementary transformations the unknowns in the system are sequentially annihilated. In other words, by making elementary transformations on the rows of the augmented matrix, this matrix is reduced to a step form.

If in the course of these transformations in this matrix we get a row with all zero elements, as the equation corresponding to this row is in the form

$$0 \cdot x_1 + 0 \cdot x_2 + \ldots + 0 \cdot x_n = 0,$$

we can reject this row.

In the case of transformation, if the matrix has a row with a non-zero element in the last column, as the equation corresponding to this row is in the form

$$0\cdot x_1 + 0\cdot x_2 + \ldots + 0\cdot x_n = b, \quad (b \neq 0)$$

and this equation has no solution, we can stop transformations and state that the given system is not adjoint.

If the number of the step-matrix is r, two cases are possible:

1. **Case.** $r = n$. In this case, the system of linear equations corresponding to the step-matrix is in the form:

$$\left.\begin{aligned} a'_{11}x_1 + a'_{12}x_2 + \ldots + a'_{1n-1}x_{n-1} + a'_{1n}x_n &= b'_1, \\ a'_{22}x_2 + \ldots + a'_{2n-1}x_{n-1} + a'_{2n}x_n &= b'_2, \\ \cdots\cdots\cdots\cdots\cdots\cdots\cdots\cdots & \\ a'_{n-1n-1}x_{n-1} + a'_{n-1n}x_n &= b'_{n-1}, \\ a'_{nn}x_n &= b'_n \end{aligned}\right\}.$$

Having found $x_n - i$ from the last equation of this system and writing it in the equation last but one, $x_{n-1} - i$, etc., continuing the process, at last we find $x_1 - i$ from the first equation. In this case, the system has a unique solution. Solve the system of linear equations

$$\begin{cases} x_2 + x_3 + x_4 = -1, \\ 2x_1 + 3x_2 - x_3 + 4x_4 = 1, \\ x_1 - 2x_2 + 3x_3 - 2x_4 = -2, \\ 3x_1 + x_2 + 2x_3 + 2x_4 = -1, \\ -2x_1 + x_2 - 4x_3 + 5x_4 = -6 \end{cases}$$

by the Gauss method. For that we write its augmented matrix:

$$A' = \left(\begin{array}{cccc|c} 0 & 1 & 1 & 1 & -1 \\ 2 & 3 & -1 & 4 & 1 \\ 1 & -2 & 3 & -2 & -2 \\ 3 & 1 & 2 & 2 & -1 \\ -2 & 1 & -4 & 5 & -6 \end{array}\right).$$

Reduce this matrix to the step form. For that at first we displace the rows, take the third row with 1 in the first column to the first row and in the first column of the obtained matrix tend to zero all the elements except the first element:

$$A' \to \left(\begin{array}{cccc|c} 1 & -2 & 3 & -2 & -2 \\ 0 & 1 & 1 & 1 & -1 \\ 2 & 3 & -1 & 4 & 1 \\ 3 & 1 & 2 & 2 & -1 \\ -2 & 1 & -4 & 5 & -6 \end{array}\right) \to \left(\begin{array}{cccc|c} 1 & -2 & 3 & -2 & -2 \\ 0 & 1 & 1 & 1 & -1 \\ 0 & 7 & -7 & 8 & 5 \\ 0 & 7 & -7 & 8 & 5 \\ 0 & -3 & 2 & 1 & -10 \end{array}\right) \to$$

At first, multiply the third row by -1 and add to the fourth row, then reject the row with zero elements, at first multiply the second row by -7 and add it to the third row, then by 3 and add it to the fourth row:

$$\to \left(\begin{array}{cccc|c} 1 & -2 & 3 & -2 & -2 \\ 0 & 1 & 1 & 1 & -1 \\ 0 & 7 & -7 & 8 & 5 \\ 0 & 0 & 0 & 0 & 0 \\ 0 & -3 & 2 & 1 & -10 \end{array}\right) \to \left(\begin{array}{ccccc} 1 & -2 & 3 & -2 & -2 \\ 0 & 1 & 1 & 1 & -1 \\ 0 & 0 & -14 & 1 & 12 \\ 0 & 0 & 5 & 4 & -13 \end{array}\right) \to$$

For reducing the last matrix to the step form, at first we could multiply the third row by 5/14, and add it to the fourth row. But in order not to obtain fractional numbers, at first we multiply the fourth row by 3 and add it to the third row. Then we multiply the third row of the obtained matrix by 5 and add it to the fourth row:

$$\to \left(\begin{array}{cccc|c} 1 & -2 & 3 & -2 & -2 \\ 0 & 1 & 1 & 1 & -1 \\ 0 & 0 & 1 & 13 & -27 \\ 0 & 0 & 5 & 4 & -13 \end{array}\right) \to \left(\begin{array}{cccc|c} 1 & -2 & 3 & -2 & -2 \\ 0 & 1 & 1 & 1 & -1 \\ 0 & 0 & 1 & 13 & -27 \\ 0 & 0 & 0 & -61 & 122 \end{array}\right).$$

Write the system of linear equations corresponding to the obtained step-matrix:

$$\begin{cases} x_1 - 2x_2 + 3x_3 - 2x_4 = -2, \\ x_2 + x_3 + x_4 = -1, \\ x_3 + 13x_4 = -27, \\ -61x_4 = 122. \end{cases}$$

From the last equation, we find $x_4 = 122:(-61) = -2$. From the third equation

$$x_3 = -27 - 13x_4 = -27 - 13(-2) = -27 + 26 = -1,$$

from the second equation

$$x_2 = -1 - x_3 - x_4 = -1 - (-1) - (-2) = -1 + 1 + 2 = 2,$$

at last from the first equation

$$x_1 = -2 + 2x_2 - 3x_3 + 2x_4 = -2 + 2\cdot 2 -$$
$$3\cdot(-1) + 2\cdot(-2) = -2 + 4 + 3 - 4 = 1.$$

So, the given system has a unique solution $x_1 = 1, x_2 = 2, x_3 = -1, x_4 = -2$.

2. **Case.** $r < n$. In this case unlike the first case, the r number of rows in the step-matrix is not equal to the n number of unknowns, it is smaller than it. This means that the number of equations in the system of linear equations corresponding to the step-matrix is less than the number of unknowns. Therefore, in this case this system may not have a unique solution at all. In the obtained system, r number unknowns are retained in the left-hand side, the remaining terms with $n - r$ number unknowns are taken to the right-hand side of equations with inverse sign. These r number unknowns are called main unknowns, the remaining $n - r$ number unknowns are called free unknowns. When giving arbitrary values to free unknowns, as in the first case appropriate values of main unknowns are uniquely determined from the system with equal number of unknowns and equations. But, as we can give infinitely many values to free unknowns, the system has infinite number of solutions. Let us solve the system of linear equations

$$\begin{cases} 3x_1 + 2x_2 + 2x_3 + 2x_4 = 2, \\ 2x_1 + 3x_2 + 2x_3 + 5x_4 = 3, \\ 9x_1 + x_2 + 4x_3 - 5x_4 = 1, \\ 2x_1 + 2x_2 + 3x_3 + 4x_4 = 5, \\ 7x_1 + x_2 + 6x_3 - x_4 = 7 \end{cases}$$

by the Gauss method. Multiply the second row of the augmented matrix by -1, add to the first row. Multiply the first row of the obtained matrix by -2, add it to the second row, multiply by -9 and add to the third row, multiply by -2 and add to the fourth row, multiply by -7, add to the fifth row:

$$\left(\begin{array}{cccc|c} 3 & 2 & 2 & 2 & 2 \\ 2 & 3 & 2 & 5 & 3 \\ 9 & 1 & 4 & -5 & 1 \\ 2 & 2 & 3 & 4 & 5 \\ 7 & 1 & 6 & -1 & 7 \end{array}\right) \to \left(\begin{array}{cccc|c} 1 & -1 & 0 & -3 & -1 \\ 2 & 3 & 2 & 5 & 3 \\ 9 & 1 & 4 & -5 & 1 \\ 2 & 2 & 3 & 4 & 5 \\ 7 & 1 & 6 & -1 & 7 \end{array}\right) \to \left(\begin{array}{cccc|c} 1 & -1 & 0 & -3 & -1 \\ 0 & 5 & 2 & 11 & 5 \\ 0 & 10 & 4 & 22 & 10 \\ 0 & 4 & 3 & 10 & 7 \\ 0 & 8 & 6 & 20 & 14 \end{array}\right) \to$$

$$\to \left(\begin{array}{cccc|c} 1 & -1 & 0 & -3 & -1 \\ 0 & 5 & 2 & 11 & 5 \\ 0 & 0 & 0 & 0 & 0 \\ 0 & 4 & 3 & 10 & 7 \\ 0 & 0 & 0 & 0 & 0 \end{array}\right) \to \left(\begin{array}{ccccc} 1 & -1 & 0 & -3 & -1 \\ 0 & 5 & 2 & 11 & 5 \\ 0 & 4 & 3 & 10 & 7 \end{array}\right) \to$$

(multiply the third row by -1 and add to the second row. Multiply the second row of the obtained matrix by -4 and add to the third row)

$$\to \left(\begin{array}{cccc|c} 1 & -1 & 0 & -3 & -1 \\ 0 & 1 & -1 & 1 & -2 \\ 0 & 4 & 3 & 10 & 7 \end{array}\right) \to \left(\begin{array}{cccc|c} 1 & -1 & 0 & -3 & -1 \\ 0 & 1 & -1 & 1 & -2 \\ 0 & 0 & 7 & 6 & 15 \end{array}\right).$$

The rank of the last matrix is 3. The rank is smaller than the number of unknowns. Write the system of linear equations appropriate to this matrix:

$$\begin{cases} x_1 - x_2 \qquad - 3x_4 = -1, \\ \qquad x_2 - x_3 + x_4 = -2, \\ \qquad\qquad 7x_3 + 6x_4 = 15. \end{cases}$$

In such a system the unknowns corresponding to the column number of non-zero first element from each row of the step-matrix, should be taken as main unknowns. In our example, x_1, x_2, and x_3 are free unknowns, x_4 is a free unknown. In the last system, we take the terms with x_4 to the right-hand side of equations:

$$\begin{cases} x_1 - x_2 = -1 + 3x_4, \\ x_2 - x_3 = -2 - x_4, \\ 7x_3 = 15 - 6x_4 \ . \end{cases} \tag{1.7}$$

If we give an arbitrary value to x_4, then x_1, x_2, x_3 may be found uniquely. For example, if we take $x_4 = 0$, we get

$$\begin{cases} x_1 - x_2 = -1, \\ x_2 - x_3 = -2, \\ 7x_3 = 15 . \end{cases}$$

Hence, we find

$$x_3 = \frac{15}{7}, \ x_2 = -2 + x_3 = -2 + \frac{15}{7} = \frac{1}{7}, \ x_1 = -1 + x_2 = -1 + \frac{1}{7} = -\frac{6}{7}$$

$$\left(x_1 = -\frac{6}{7}, \ x_2 = \frac{1}{7}, \ x_3 = \frac{15}{7}, \ x_4 = 0 \right).$$

If in (1.7) we take $x_4 = 1$, then having solved the obtained system

$$\begin{cases} x_1 - x_2 = 2, \\ x_2 - x_3 = -3, \\ 7x_3 = 9 \end{cases}$$

we find

$$x_3 = \frac{9}{7}, \ x_2 = -3 + x_3 = -3 + \frac{9}{7} = -\frac{12}{7}, \ x_1 = 2 + x_2 = 2 - \frac{12}{7} = \frac{2}{7}$$

$$\left(x_1 = \frac{2}{7}, \ x_2 = -\frac{12}{7}, \ x_3 = \frac{9}{7}, \ x_4 = 1 \right).$$

The solutions obtained by giving such values to the free unknown are called particular solutions of the system. They are infinitely many.

If in (2) we take $x_4 = c$ and solve the system, we find

$$x_3 = \frac{15}{7} - \frac{6}{7}c, \ x_2 = -2 - c + x_3 = -2 - c + \frac{15}{7} - \frac{6}{7}c = \frac{1}{7} - \frac{13}{7}c,$$

$$x_1 = -1 + 3c + x_2 = -1 + 3c + \frac{1}{7} - \frac{13}{7}c = -\frac{6}{7} + \frac{8}{7}c.$$

The found formula

$$x_1 = -\frac{6}{7} + \frac{8}{7}c, \quad x_2 = \frac{1}{7} - \frac{13}{7}c, \quad x_3 = \frac{15}{7} - \frac{6}{7}c, \quad x_4 = c$$

is called a generalized solution of the system under consideration. Each particular solution may be obtained from general solution. If in the formulas of general solution we take $c = 0$, we get the first, for $c = 1$ we get the second particular solution.

Now let us consider the system with equal number of unknowns and equations:

$$\begin{cases} a_{11}x_1 + a_{12}x_2 + \ldots + a_{1n}x_n = b_1, \\ a_{21}x_1 + a_{22}x_2 + \ldots + a_{2n}x_n = b_2, \\ \cdot \quad \cdot \quad \cdot \quad \cdot \quad \cdot \quad \cdot \quad \cdot \quad \cdot \quad \cdot \quad \cdot \quad \cdot \quad \cdot \\ a_{n1}x_1 + a_{n2}x_2 + \ldots + a_{nn}x_n = b_n. \end{cases} \tag{1.8}$$

From the coefficients of this system we form the following determinants:

$$\Delta = \begin{vmatrix} a_{11} & a_{12} & \cdot & \cdot & \cdot & a_{1n} \\ a_{21} & a_{22} & \cdot & \cdot & \cdot & a_{2n} \\ \cdot & \cdot & \cdot & \cdot & \cdot & \cdot \\ a_{n1} & a_{n2} & \cdot & \cdot & \cdot & a_{nn} \end{vmatrix}, \qquad \Delta_1 = \begin{vmatrix} b_1 & a_{12} & \cdot & \cdot & \cdot & a_{1n} \\ b_2 & a_{22} & \cdot & \cdot & \cdot & a_{2n} \\ \cdot & \cdot & \cdot & \cdot & \cdot & \cdot \\ b_n & a_{n2} & \cdot & \cdot & \cdot & a_{nn} \end{vmatrix},$$

$$\Delta_2 = \begin{vmatrix} a_{11} & b_1 & \cdot & \cdot & \cdot & a_{1n} \\ a_{21} & b_2 & \cdot & \cdot & \cdot & a_{2n} \\ \cdot & \cdot & \cdot & \cdot & \cdot & \cdot \\ a_{n1} & b_n & \cdot & \cdot & \cdot & a_{nn} \end{vmatrix}, \ \ldots, \qquad \Delta_n = \begin{vmatrix} a_{11} & a_{12} & \cdot & \cdot & \cdot & b_1 \\ a_{21} & a_{22} & \cdot & \cdot & \cdot & b_2 \\ \cdot & \cdot & \cdot & \cdot & \cdot & \cdot \\ a_{n1} & a_{n2} & \cdot & \cdot & \cdot & b_n \end{vmatrix}.$$

Δ is called the main determinant of (1.8), $\Delta_1, \Delta_2, \ldots, \Delta_n$ auxiliary determinants. The Δ_i determinant is obtained from Δ replacing its first column by a column consisting of free terms of eq 1.8 ($i = 1, 2, \ldots, n$).

Theorem (*Kramer*). For $\Delta \neq 0$, (3) has a unique solution and for this solution the following formula is valid

$$x_1 = \frac{\Delta_1}{\Delta}, \quad x_2 = \frac{\Delta_2}{\Delta}, \quad \ldots, \quad x_n = \frac{\Delta_n}{\Delta}. \tag{1.9}$$

Equation 1.9 is said to be Kramer's formula.

In eq 1.8, when $b = b_2 = ... = b_n = 0$, the obtained system is said to be a system of linear equations. Homogeneous system of linear equations always has the trivial solution $x_1 = x_2 = ... = x_n = 0$. In order homogeneous system have non-trivial solution, the main determinant of this system be equal to zero.

Problems to be solved in auditorium

Problem 30. Solve the following system of linear equations by the Kramer method:

$$1)\begin{cases} 3x-4y=-6, \\ 3x+4y=18; \end{cases} \qquad 2)\begin{cases} 7x+2y+3z=15, \\ 5x-3y+2z=15, \\ 10x-11y+5z=36; \end{cases}$$

$$3)\begin{cases} 4x_1+4x_2+5x_3+5x_4=0, \\ 2x_1 \qquad +3x_3-x_4=10, \\ x_1 \quad +x_2 \ -5x_3 \qquad =-10, \\ \qquad 3x_2 \ +2x_3 \qquad =1; \end{cases}; \qquad 4)\begin{cases} 2x_1+2x_2-x_3 \quad +x_4 \ =4, \\ 4x_1+3x_2-x_3+2x_4=6, \\ 8x_1+5x_2-3x_3+4x_4=12, \\ 3x_1+3x_2-2x_3+2x_4=6. \end{cases}$$

Solution of 3): At first calculate the determinant of the system:

$$\Delta = \begin{vmatrix} 4 & 4 & 5 & 5 \\ 2 & 0 & 3 & -1 \\ 1 & 1 & -5 & 0 \\ 0 & 3 & 2 & 0 \end{vmatrix} = \begin{vmatrix} 14 & 4 & 20 & 0 \\ 2 & 0 & 3 & -1 \\ 1 & 1 & -5 & 0 \\ 0 & 3 & 2 & 0 \end{vmatrix} = -1\cdot(-1)^{2+4}\cdot \begin{vmatrix} 14 & 4 & 20 \\ 1 & 1 & -5 \\ 0 & 3 & 2 \end{vmatrix} =$$
$$= -(28+60-8+210) = -290.$$

As $\Delta \neq 0$, the system has a unique solution and this solution may be found by the Kramer formulas. For that calculate the auxiliary determinant:

$$\Delta_1 = \begin{vmatrix} 0 & 4 & 5 & 5 \\ 10 & 0 & 3 & -1 \\ -10 & 1 & -5 & 0 \\ 1 & 3 & 2 & 0 \end{vmatrix} = \begin{vmatrix} 50 & 4 & 20 & 0 \\ 10 & 0 & 3 & -1 \\ -10 & 1 & -5 & 0 \\ 1 & 3 & 2 & 0 \end{vmatrix} = -\begin{vmatrix} 50 & 4 & 20 \\ -10 & 1 & -5 \\ 1 & 3 & 2 \end{vmatrix} =$$
$$= -(100-20-600-20+80+750) = -290,$$

$$\Delta_2 = \begin{vmatrix} 4 & 0 & 5 & 5 \\ 2 & 10 & 3 & -1 \\ 1 & -10 & -5 & 0 \\ 0 & 1 & 2 & 0 \end{vmatrix} = \begin{vmatrix} 14 & 50 & 20 & 0 \\ 2 & 10 & 3 & -1 \\ 1 & -10 & -5 & 0 \\ 0 & 1 & 2 & 0 \end{vmatrix} =$$

$$= -1 \cdot \begin{vmatrix} 14 & 50 & 20 \\ 1 & -10 & -5 \\ 0 & 1 & 2 \end{vmatrix} = -(-280 + 20 - 100 + 70) = 290,$$

$$\Delta_3 = \begin{vmatrix} 4 & 4 & 0 & 5 \\ 2 & 0 & 10 & -1 \\ 1 & 1 & -10 & 0 \\ 0 & 3 & 1 & 0 \end{vmatrix} = \begin{vmatrix} 14 & 4 & 50 & 0 \\ 2 & 0 & 10 & -1 \\ 1 & 1 & -10 & 0 \\ 0 & 3 & 1 & 0 \end{vmatrix} = -1 \cdot \begin{vmatrix} 14 & 4 & 50 \\ 1 & 1 & -10 \\ 0 & 3 & 1 \end{vmatrix} =$$

$$= -(14 + 150 - 4 + 420) = -580,$$

$$\Delta_4 = \begin{vmatrix} 4 & 4 & 5 & 0 \\ 2 & 0 & 3 & 10 \\ 1 & 1 & -5 & -10 \\ 0 & 3 & 2 & 1 \end{vmatrix} = \begin{vmatrix} 4 & 4 & 5 & 0 \\ 2 & -30 & -17 & 0 \\ 1 & 31 & 15 & 0 \\ 0 & 3 & 2 & 1 \end{vmatrix} = \begin{vmatrix} 4 & 4 & 5 \\ 2 & -30 & -17 \\ 1 & 31 & 15 \end{vmatrix} =$$

$$= \begin{vmatrix} 0 & -120 & -55 \\ 0 & -92 & -47 \\ 1 & 31 & 15 \end{vmatrix} = \begin{vmatrix} -120 & -55 \\ -92 & -47 \end{vmatrix} = 5640 - 5060 = 580.$$

By the Kramer formula we find:

$$x_1 = \frac{\Delta_1}{\Delta} = \frac{-290}{-290} = 1; \qquad x_2 = \frac{\Delta_2}{\Delta} = \frac{290}{-290} = -1;$$

$$x_3 = \frac{\Delta_3}{\Delta} = \frac{-580}{-290} = 2; \qquad x_4 = \frac{\Delta_4}{\Delta} = \frac{580}{-290} = -2.$$

Answer: 1) $x = 2$, $u = 3$; 2) $x = 2$, $u = -1$, $z = 1$;
4) $x_1 = 1$, $x_2 = 1$, $x_3 = -1$, $x_4 = -1$.

Problem 31. Study compatibility of the system of linear equations, if it is compatible find, its general solution:

1) $\begin{cases} 3x_1 - 2x_2 - 5x_3 + x_4 = 3, \\ 2x_1 - 3x_2 + x_3 + 5x_4 = -3, \\ x_1 + 2x_2 \quad -4x_4 = -3, \\ x_1 - x_2 - 4x_3 + 9x_4 = 22; \end{cases}$ 2) $\begin{cases} 3x_1 - 5x_2 + 2x_3 + 4x_4 = 2, \\ 7x_1 - 4x_2 + x_3 + 3x_4 = 5, \\ 5x_1 + 7x_2 - 4x_3 - 6x_4 = 3; \end{cases}$

3) $\begin{cases} 2x_1 - x_2 + x_3 + 2x_4 + 3x_5 = 2, \\ 6x_1 - 3x_2 + 2x_3 + 4x_4 + 5x_5 = 3, \\ 6x_1 - 3x_2 + 4x_3 + 8x_4 + 13x_5 = 9, \\ 4x_1 - 2x_2 + x_3 + x_4 + 2x_5 = 4. \end{cases}$

Answer: 1) The system has a unique solution: $x_1 = -1$, $x_2 = 3$, $x_3 = -2$, $x_4 = 2$; 2) the system is not compatible; 3) the system is compatible and has infinitely many solutions, general solution:

$$x_1 = \frac{1}{2} \cdot (-1 + c_1 + c_2),\ x_2 = c_1,\ x_3 = 9 - 4c_2,\ x_4 = -3,$$
$$x_5 = c_2 \ (c_1,\ c_2 \text{ are arbitrary numbers}).$$

Problem 32. Solve the system of homogeneous linear equations:

1) $\begin{cases} x_1 + x_2 + x_3 + x_4 + x_5 = 0, \\ x_1 - x_2 + 2x_3 - 2x_4 + 3x_5 = 0, \\ x_1 + x_2 + 4x_3 + 4x_4 + 9x_5 = 0, \\ x_1 - x_2 + 8x_3 - 8x_4 + 27x_5 = 0, \\ x_1 + x_2 + 16x_3 + 16x_4 + 81x_5 = 0; \end{cases}$ 2) $\begin{cases} x_1 + 3x_2 + 2x_3 = 0, \\ 2x_1 - x_2 + 3x_3 = 0, \\ 3x_1 - 5x_2 + 4x_3 = 0, \\ x_1 + 17x_2 + 4x_3 = 0; \end{cases}$

3) $\begin{cases} x_1 + x_2 \quad -3x_4 - x_5 = 0, \\ x_1 - x_2 + x_3 - x_4 \quad = 0 \\ 4x_1 - 2x_2 + 6x_3 + 3x_4 - 4x_5 = 0, \\ 2x_1 + 4x_2 - 2x_3 + 4x_4 - 7x_5 = 0. \end{cases}$

Solution of 2): As the system is homogeneous, $x_1 = x_2 = x_3 = 0$ has a trivial solution. By the Gauss method, we study if the system has a nontrivial solution. At first, write the main matrix of the system and reduce it to the step form:

$$\begin{pmatrix} 1 & 3 & 2 \\ 2 & -1 & 3 \\ 3 & -5 & 4 \\ 1 & 17 & 4 \end{pmatrix} \to \begin{pmatrix} 1 & 3 & 2 \\ 0 & -7 & -1 \\ 0 & -14 & -2 \\ 0 & 14 & 2 \end{pmatrix} \to \begin{pmatrix} 1 & 3 & 2 \\ 0 & -7 & -1 \\ 0 & 0 & 0 \\ 0 & 0 & 0 \end{pmatrix} \to \begin{pmatrix} 1 & 3 & 2 \\ 0 & -7 & -1 \end{pmatrix}.$$

Write the linear homogeneous system with the main last matrix:

$$\begin{cases} x_1 + 3x_2 + 2x_3 = 0, \\ -7x_2 - x_3 = 0 \end{cases} \Rightarrow \begin{cases} x_1 + 3x_2 = -2x_3, \\ -7x_2 = x_3 \end{cases} \Rightarrow$$

$$\Rightarrow \begin{cases} x_2 = -\dfrac{1}{7}x_3, \\ x_1 = -2x_3 - 3x_2 \end{cases} \Rightarrow \begin{cases} x_2 = -\dfrac{1}{7}x_3, \\ x_1 = -\dfrac{11}{7}x_3. \end{cases}$$

If we take $x_3 = c$, we can write the general solution of the given system in the form $x_1 = -\frac{11}{7} \cdot c$, $x_2 = -\frac{1}{7} \cdot c$, $x_3 = c$ (here c is an arbitrary number).

Answer: 1) the system has only a trivial solution: $x_1 = x_2 = x_3 = x_4 = x_5 = 0$; 2) the system has infinitely many solutions, general solution: $x_1 = (7/6) \cdot c_5 - c_3$, $x_2 = (5/6) \cdot c_5 + c_3$, $x_3 = c_3$, $x_4 = (1/3) \cdot c_5$, $x_5 = c_5$ (here c_3, c_5 are arbitrary numbers).

Home tasks

Problem 33. Solve the systems by the Kramer method:

1) $\begin{cases} x_1 + x_2 + x_3 = 6, \\ x_1 + x_2 - x_3 = 0, \\ 2x_1 - 2x_2 + 3x_3 = 7; \end{cases}$ 2) $\begin{cases} 3x_1 - 2x_2 + x_3 = -14, \\ x_1 + 4x_2 = 5, \\ 5x_1 - 3x_3 = -12; \end{cases}$

3) $\begin{cases} 2x_1 + 3x_2 + 11x_3 + 5x_4 = 2, \\ x_1 + x_2 + 5x_3 + 2x_4 = 1, \\ 2x_1 + x_2 + 3x_3 + 2x_4 = -3, \\ x_1 + x_2 + 3x_3 + 4x_4 = -3. \end{cases}$

Answer: 1) $x_1 = 1$, $x_2 = 2$, $x_3 = 3$; 2) $x_1 = -3$, $x_2 = 2$, $x_3 = -1$; 3) $x_1 = -2$, $x_2 = 0$, $x_3 = 1$, $x_4 = -1$.

Problem 34. Study compatibility of the system of linear equations. If it is compatible, find its general solution:

$$1)\ \begin{cases} x_1 + x_2 - 6x_3 - 4x_4 = 6, \\ 3x_1 - x_2 - 6x_3 - 4x_4 = 2, \\ 2x_1 + 3x_2 + 9x_3 + 2x_4 = 6, \\ 3x_1 + 2x_2 + 3x_3 + 8x_4 = -7; \end{cases} \qquad 2)\ \begin{cases} x_1 + x_2 + 3x_3 - 2x_4 + 3x_5 = 1, \\ 2x_1 + 2x_2 + 4x_3 - x_4 + 3x_5 = 2, \\ 3x_1 + 3x_2 + 5x_3 - 2x_4 + 3x_5 = 1, \\ 2x_1 + 2x_2 + 8x_3 - 3x_4 + 9x_5 = 2; \end{cases}$$

$$3)\ \begin{cases} 12x_1 + 14x_2 - 15x_3 + 24x_4 + 27x_5 = 5, \\ 16x_1 + 18x_2 - 22x_3 + 29x_4 + 37x_5 = 8, \\ 18x_1 + 20x_2 - 21x_3 + 32x_4 + 41x_5 = 9, \\ 10x_1 + 12x_2 - 16x_3 + 20x_4 + 23x_5 = 4. \end{cases}$$

Answer: 1) the system is compatible, has a unique solution: $x_1 = 0$, $x_2 = 2$, $x_3 = 1/3$, $x_4 = -3/2$; 2) the system is not compatible; 3) the system is compatible, has infinitely many solutions, general solution:

$$x_1 = \frac{40}{18} - \frac{53}{18}c, \quad x_2 = -\frac{5}{3} + \frac{5}{6}c, \quad x_3 = -\frac{1}{9} + \frac{2}{9}c,$$
$$x_4 = 0, \quad x_5 = c.$$

Problem 35. Solve the homogeneous system of linear equations:

$$1)\ \begin{cases} x_1 - 2x_2 + x_3 + x_4 - x_5 = 0, \\ 2x_1 + x_2 - x_3 - x_4 + x_5 = 0, \\ x_1 + 7x_2 - 5x_3 - 5x_4 + 5x_5 = 0, \\ 3x_1 - x_2 - 2x_3 + x_4 - x_5 = 0; \end{cases} \qquad 2)\ \begin{cases} 3x_1 + 4x_2 - 5x_3 + 7x_4 = 0, \\ 2x_1 - 3x_2 + 3x_3 - 2x_4 = 0, \\ 4x_1 + 11x_2 - 13x_3 + 16x_4 = 0, \\ 7x_1 - 2x_2 + x_3 + 3x_4 = 0. \end{cases}$$

Answer:

1) $x_1 = x_2 = x_3 = 0, \quad x_4 = x_5;$

2) $x_1 = \dfrac{3c_3 - 13c_4}{17}, \quad x_2 = \dfrac{19c_3 - 20c_4}{17}$, here c_3, c_4 are arbitrary numbers.

$x_3 = c_3, \quad x_4 = c_4.$

1.6 LINEAR OPERATIONS ON VECTORS: BASIS VECTORS IN PLANE AND SPACE

A directed straight line is called a vector. The length of a straight line indicating a vector is called modulus of vector or its length. Vectors are denoted either by two capital letters of the Latin alphabet, for example, $\vec{AB}$, $\vec{CD}$, etc. or by one small letter of the Latin alphabet, for example, as $\vec{a}$, $\vec{b}$, $\vec{c}$, etc. For giving a vector, its length and direction should be known. A vector whose origin and end coincides, is called a zero vector and is denoted by $\vec{0}$.

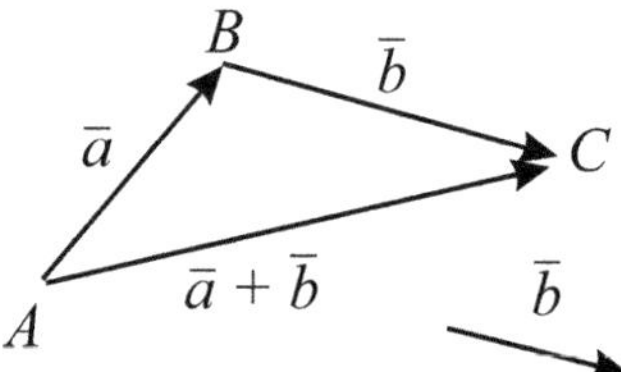

It is possible to put together two vectors, subtract one vector from another one, to multiply a vector by any number. These are called linear operations on vectors.

For putting together the vectors $\vec{a}$ and $\vec{b}$, it is necessary to draw the vector $\vec{b}=\vec{BC}$ from the end of the vector $\vec{a}=\vec{AB}$ and connect A that is the beginning of $\vec{a}$ and the C, the end of $\vec{b}$. The vector $\vec{AC}$ is called the sum of the vectors $\vec{a}$ and $\vec{b}$:

$$\vec{AC}=\vec{a}+\vec{b}.$$

Product of the vector $\vec{a}$ and a real number λ is a vector denoted by $\lambda\vec{a}$ or $\vec{a}\lambda$ and determined by the following rule:

1. $\left|\lambda\vec{a}\right|=|\lambda|\cdot\left|\vec{a}\right|$;
2. for $\lambda>0$, the vectors $\vec{a}$ and $\lambda\vec{a}$ are same-directional, for $\lambda<0$ they are opposite directional.

The vector $\vec{a}\cdot(-1)$ denoted as $-\vec{a}$.

To subtract the vector $\vec{b}$ from the vector $\vec{a}$ means to put together the vectors $\vec{a}$ and $-\vec{b}$.

Addition of vectors and multiplication by a number has the following properties:

1. $\vec{a}+\vec{0}=\vec{a}$;
2. $\vec{a}+\vec{b}=\vec{b}+\vec{a}$;
3. $\vec{a}+\left(\vec{b}+\vec{c}\right)=\left(\vec{a}+\vec{b}\right)+\vec{c}$;
4. $0\cdot\vec{a}=\vec{0}$;
5. $\lambda\cdot\vec{0}=\vec{0}$;
6. $(\lambda_1\lambda_2)\vec{a}=\lambda_1\left(\lambda_2\vec{a}\right)$;
7. $\lambda\left(\vec{a}+\vec{b}\right)=\lambda\vec{a}+\lambda\vec{b}$;
8. $(\lambda_1+\lambda_2)\vec{a}=\lambda_1\vec{a}+\lambda_2\vec{a}$.

Definition. The vectors located on the same straight line or on parallel straight lines are called collinear vectors.

It is clear that collinear vectors are either same directional or opposite directional. Therefore, when $\vec{a}$ and $\vec{b}$ are collinear, there always may be found the numbers λ and μ that $\vec{a}=\lambda\vec{b}$, $\vec{b}=\mu\vec{a}$.

Definition. Two non-collinear vectors successively taken on a plane are called basis vectors on this plane. Usually, the basis vectors on a plane are denoted by $\vec{e_1}$, $\vec{e_2}$.

Theorem. Arbitrary $\vec{a}$ vector on a plane may be uniquely separated on any basis $\vec{e_1}$, $\vec{e_2}$ on this plane in the form

$$\vec{a} = \lambda_1 \vec{e_1} + \lambda_2 \vec{e_2} \cdot \tag{1.10}$$

The numbers λ_1 and λ_2 are called the coordinates of the vector $\vec{a}$ with respect to the basis $\vec{e_1}, \vec{e_2}$ on a plane. When basis vectors are relatively perpendicular and the length of each of them equals a unit, it is called an orthonormal basis.

Basis vectors on a plane are written as $\vec{i}, \vec{j}$ $\left(\left|\vec{i}\right| = \left|\vec{j}\right| = 1, \left(\widehat{\vec{i}, \vec{j}}\right) = 90^\circ \right)$, expansion of arbitrary $\vec{a}$ vector in such a basis is written as

$$\vec{a} = a_x \vec{i} + a_y \vec{j} \cdot$$

By $\vec{a}\,(a_x, a_y)$ we denote that the numbers a_x and a_u are the coordinates of $\vec{a}$ vector. The length of $\vec{a}\,(a_x,\ a_y)$ vector may be found by the formula

$$\left|\vec{a}\right| = \sqrt{a_x^2 + a_y^2}\ .$$

Definition. Three vectors arranged on the same plane or parallel planes in space, are called coplanar vectors.

Definition. If in space, for the vectors $\vec{a}, \vec{b}, \vec{c}$ there are three numbers $\lambda_1, \lambda_2, \lambda_3$ one of which is non-zero, such that

$$\lambda_1 \vec{a} + \lambda_2 \vec{b} + \lambda_3 \vec{c} = \vec{0},$$

then $\vec{a}, \vec{b}, \vec{c}$ is called the system of linearly dependent vectors.

Theorem. Necessary and sufficient condition for three vectors in a space be coplanar is their linear dependence.

Theorem. Noncoplanar three vectors taken with certain succession in space are called basis vectors in this space.

Arbitrary basis in a space is denoted as $\vec{e_1}, \vec{e_2}, \vec{e_3}$, an orthonormal basis as

$$\vec{i},\ \vec{j},\ \vec{k},\ \left(\left|\vec{i}\right| = \left|\vec{j}\right| = \left|\vec{k}\right| = 1,\quad \left(\widehat{\vec{i},\vec{j}}\right) = \left(\widehat{\vec{i},\vec{k}}\right) = \left(\widehat{\vec{j},\vec{k}}\right) = 90^\circ\right).$$

We can write expansion of the vector $\vec{a}$ in the basis $\vec{e_1}$, $\vec{e_2}$, $\vec{e_3}$ in the space in the form

$$\vec{a} = \lambda_1 \vec{e_1} + \lambda_2 \vec{e_2} + \lambda_3 \vec{e_3}, \tag{1.11}$$

expansion in $\vec{i}$, $\vec{j}$, $\vec{k}$ orthonormal basis in the form

$$\vec{a} = a_x \vec{i} + a_y \vec{j} + a_z \vec{k}\cdot$$

In space, the length of the vector $\vec{a}(a_x, a_y, a_z)$ is found by the formula

$$\left|\vec{a}\right| = \sqrt{a_x^2 + a_y^2 + a_z^2}.$$

When putting together two vectors their appropriate coordinates are put together. When multiplying a vector by a number, all its components are multiplied by this number. Coordinates in orthonormal basis coincide with coordinates in the rectangular coordinates system.

$\overrightarrow{M_1 M_2}(x_2 - x_1,\ y_2 - y_1,\ z_2 - z_1)$ is a vector connecting the points $M_1(x_1, y_1, z_1)$ and $M_2(x_2, y_2, z_2)$.

Necessary and sufficient condition for the vectors $\vec{a}(a_x, a_y)$ and $\vec{b}(b_x, b_y)$ on a plane be collinear is proportionality of their appropriate coordinates:

$$\frac{a_x}{b_x} = \frac{a_y}{b_y}.$$

In order to know if the vectors $\vec{a}(a_x, a_y)$ and $\vec{b}(b_x, b_y)$ with known coordinates on a plane form a basis, it is necessary to calculate the rank of one of the second-order matrices made of the coordinates of these vectors

$$A_1 = \begin{pmatrix} a_x & b_x \\ a_y & b_y \end{pmatrix} \text{ or } A_2 = \begin{pmatrix} a_x & a_y \\ b_x & b_y \end{pmatrix}.$$

When the rank of any of these matrices equals 2, $\vec{a}$ and $\vec{b}$ are basis vectors. For example, for the vectors $\vec{a}$ (1; 2), $\vec{b}$ (3; 4)

$$A_1 = \begin{pmatrix} 1 & 3 \\ 2 & 4 \end{pmatrix}, \ \det A_1 = \begin{vmatrix} 1 & 3 \\ 2 & 4 \end{vmatrix} = 4 - 6 = -2 \neq 0 \Rightarrow rank\ A_1 = 2$$

$\vec{a}$ and $\vec{b}$ are basis vectors.

For the vectors $\vec{a}(a_x, a_y, a_z)$, $\vec{b}(b_x, b_y, b_z)$, $\vec{c}(c_x, c_y, c_z)$ to form a basis in space, the rank of the matrix

$$A_1 = \begin{pmatrix} a_x & b_x & c_x \\ a_y & b_y & c_y \\ a_z & b_z & c_z \end{pmatrix} \text{ or } A_2 = \begin{pmatrix} a_x & a_y & a_z \\ b_x & b_y & b_z \\ c_x & c_y & c_z \end{pmatrix}$$

to be equal to 3.

For finding coordinates of the vector $\vec{a}$ in the given basis vectors, expressing equality (1) by coordinates when it is on a plane and (2) when it is on the space, the obtained linear system of equations with two or three unknowns should be solved. For example, for finding the coordinates x, y of the vector $\vec{a}$ (3; 1) in basis vectors $\vec{e_1}$ (2; 4), $\vec{e_2}$ (5; −3) it is necessary to solve the following system of linear equations

$$\vec{a} = x\vec{e_1} + y\vec{e_2} \Rightarrow (3, 1) = x \cdot (2; 4) + y \cdot (5; -3) \Rightarrow \begin{cases} 2x + 5y = 3, \\ 4x - 3y - 1 \end{cases}.$$

Problems to be solved in auditorium

Problem 36. For $A(1; 3; 2)$ and $B(5; 8; -1)$, find the vector $\vec{a} = \vec{AB}$ and its length.

Answer: $\vec{a}\,(4; 5; -3)$, $\quad \left|\vec{a}\right| = 5\sqrt{2}$.

Problem 37. Find the length of the vector $\vec{a}=4\vec{i}+2\vec{j}-4\vec{k}$ and the cosines of direction angles.

***Guideline*:** For the cosines of direction angles α, β, γ formed by the vectors $\vec{a}=a_x\vec{i}+a_y\vec{j}+a_z\vec{k}$ and the axes *OX, OY, OZ* use the formula:

$$\cos\alpha=\frac{a_x}{|\vec{a}|},\qquad \cos\beta=\frac{a_y}{|\vec{a}|},\qquad \cos\gamma=\frac{a_z}{|\vec{a}|}.$$

Answer: $|\vec{a}|=6,\ \ \cos\alpha=\frac{2}{3},\ \ \cos\beta=\frac{1}{3},\ \ \cos\gamma=-\frac{2}{3}.$

Problem 38. We are given the vectors

$$\vec{AB}=\vec{e_1}+2\vec{e_2},\ \vec{BC}=-4\vec{e_1}-\vec{e_2},\ \vec{CD}=-5\vec{e_1}-3\vec{e_2}\,.$$

Prove that *ABCD* is a trapezoid.

Solution. For the quadrangle *ABCD* to be a trapezoid, its two opposite sides to be parallel, that is, *AB*||*CD* or *BC*||*AD*. This means that for *ABCD* to be a trapezoid, either $\vec{AB}$ and $\vec{CD}$, or $\vec{BC}$ and $\vec{AD}$ to be collinear.

As their appropriate coordinates are not proportional, $\vec{AB}$ (1; 2) and $\vec{CD}$ (−5; −3) are not collinear.

We must verify if $\vec{BC}$ and $\vec{AD}$ are collinear. The coordinates of the vector $\vec{AD}$ are not known. We can consider the vector $\vec{AD}$ as the sum of the vectors $\vec{AB}$, $\vec{BC}$, and $\vec{CD}$. Therefore we get:

$$\vec{AD}=\vec{AB}+\vec{BC}+\vec{CD}=\left(\vec{e_1}+2\vec{e_2}\right)+$$
$$+\left(-4\vec{e_1}-\vec{e_2}\right)+\left(-5\vec{e_1}-3\vec{e_2}\right)=-8\vec{e_1}-2\vec{e_2}.$$

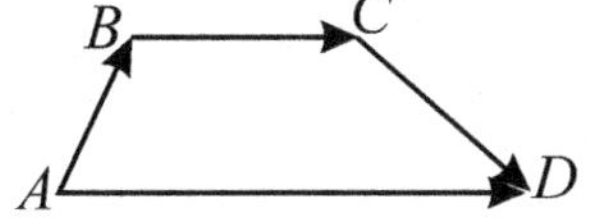

As the coordinates of the vectors $\overrightarrow{BC}$ (−4; −1) and $\overrightarrow{AD}$ (−8; −2) are proportional, they are collinear $\left(\frac{-4}{-8} = \frac{-1}{-2}\right)$. Thus, $ABCD$ is a trapezoid.

Problem 39. Prove that for arbitrary vectors $\vec{a}$, $\vec{b}$, $\vec{c}$ the vectors $\vec{a} + \vec{b}$, $\vec{b} + \vec{c}$, and $\vec{c} - \vec{a}$ are coplanar.

Problem 40. The vectors $\vec{e_1}$, $\vec{e_2}$ and $\vec{a}$ are given on a plane. Show that the vectors $\vec{e_1}$ and $\vec{e_2}$ form a basis and write the expansion of the vector $\vec{a}$ in this basis:

1) $\vec{e_1}$ (−1; 2), $\vec{e_2}$ (2; 1), $\vec{a}$ (0; −2);

2) $\vec{e_1}$ (1; 1), $\vec{e_2}$ (1; −3), $\vec{a}$ (3; −1).

Answer: 1) $\vec{a} = -\frac{4}{5}\vec{e_1} - \frac{2}{5}\vec{e_2}$; 2) $\vec{a} = 2\vec{e_1} + \vec{e_2}$.

Problem 41. The vectors $\vec{e_1}$, $\vec{e_2}$, $\vec{e_3}$, and $\vec{a}$ are given on a space. Determine if the vectors $\vec{e_1}$, $\vec{e_2}$, $\vec{e_3}$ form a basis. If they form a basis, then write the expansion of the vector $\vec{a}$ in this basis:

1) $\vec{e_1}$ (1; 0; 0), $\vec{e_2}$ (1; 1; 0), $\vec{e_3}$ (1; 1; 1), $\vec{a}$ (−2; 0; −1);

2) $\vec{e_1}$ (1; 2; 3), $\vec{e_2}$ (−2; −4; 4), $\vec{e_3}$ (−1; −2; 5), $\vec{a}$ (3; 4; 5);

3) $\vec{e_1}$ (1; 1; −1), $\vec{e_2}$ (2; 1; 0), $\vec{e_3}$ (3; 2;1), $\vec{a}$ (2; −1;3).

Solution of 3). Write a matrix whose rows were made of the coordinates of the vectors $\vec{e_1}$, $\vec{e_2}$, $\vec{e_3}$:

$$C = \begin{pmatrix} 1 & 1 & -1 \\ 2 & 1 & 0 \\ 3 & 2 & 1 \end{pmatrix}.$$

As det $C = -2 \neq 0$, rank $C = 3$. So, the vectors $\vec{e_1}$, $\vec{e_2}$, $\vec{e_3}$ make a basis.

Find now the expansion of the vector $\vec{a}$ (2; −1; −3) in this basis. If we denote the coordinates of $\vec{a}$ in the basis $\vec{e_1}$, $\vec{e_2}$, $\vec{e_3}$ by x_1, x_2, x_3, we can write $\vec{a} = x_1\vec{e_1} + x_2\vec{e_2} + x_3\vec{e_3}$. Express the last equality by coordinates:

$$(2; -1; 3) = x_1 \cdot (1; 1; -1) + x_2 \cdot (2; 1; 0) + x_3(3; 2; 1) \Rightarrow$$
$$\Rightarrow (2; -1; 3) = (x_1; x_1; -x_1) + (2x_2; x_2; 0) + (3x_3; 2x_3; x_3) \Rightarrow$$
$$\Rightarrow (2;-1;3)=(x_1+2x_2+3x_3;\ x_1+x_2+2x_3;\ -x_1+x_3).$$

Hence we get the following system of linear equations with respect to x_1, x_2, and x_3:

$$\begin{cases} x_1 + 2x_2 + 3x_3 = 2, \\ x_1 + x_2 + 2x_3 = -1, \\ -x_1 \qquad + x_3 = 3. \end{cases}$$

Having solved the system, we find $x_1 = -\frac{7}{2}$, $x_2 = \frac{7}{2}$, $x_3 = -\frac{1}{2}$. So, we get the expansion of the vector $\vec{a}$ in the basis $\vec{e_1}$, $\vec{e_2}$, $\vec{e_3}$:

$$\vec{a} = -\frac{7}{2}\vec{e_1} + \frac{7}{2}\vec{e_2} - \frac{1}{2}\vec{e_3}.$$

Answer: 1) $\vec{a} = -2\vec{e_1} + \vec{e_2} - \vec{e_3}$; 2) they do not form a basis.

Home tasks

Problem 42. Find the length and the cosines of direction angles of the vector $\vec{a} = 20\vec{i} + 30\vec{j} - 60\vec{k}$.

Answer: $|\vec{a}| = 70$; $\cos\alpha = \frac{2}{7}$; $\cos\beta = \frac{3}{7}$; $\cos\gamma = -\frac{6}{7}$.

Problem 43. Find the lengths of diagonals of parallelograms constructed on the vectors $\vec{OA} = \vec{i} + \vec{j}$ and $\vec{OB} = -3\vec{j} + \vec{k}$.

Answer: $\vec{OC} = \vec{i} - 2\vec{j} + \vec{k}$, $|\vec{OC}| = \sqrt{6}$; $\vec{AB} = -\vec{i} - 4\vec{j} + \vec{k}$, $|\vec{AB}| = 3\sqrt{2}$.

Problem 44. Noncoplanar three vectors $\vec{a}$, $\vec{b}$, $\vec{c}$ are given. Prove that the vectors $\vec{a} + 2\vec{b} - \vec{c}$, $3\vec{a} - \vec{b} + \vec{c}$, $-\vec{a} + 5\vec{b} - 3\vec{c}$ are coplanar.

Problem 45. Show that the vectors $\vec{e_1}$ and $\vec{e_2}$ form a basis on a plane. Write the expansion of the vector $\vec{a}$ in the basis $\vec{e_1}$, $\vec{e_2}$:

1) $\vec{e_1}$ $(-3; 2)$, $\vec{e_2}$ $(2; -1)$, $\vec{a}$ $(-1; -1)$;

2) $\vec{e_1}$ $(1; 4)$, $\vec{e_2}$ $(-3; -5)$, $\vec{a}$ $(2; 15)$.

Answer: 1) $\vec{a} = -3\vec{e_1} - 5\vec{e_2}$; 2) $\vec{a} = 5\vec{e_1} + \vec{e_2}$.

Problem 46. In space we are given the vectors $\vec{e_1}$, $\vec{e_2}$, $\vec{e_3}$ Verify if these vectors form a basis. If they form a basis, then find the expansion of the vector $\vec{a}$ $(6; 12; -1)$ in this basis:

1) $\vec{e_1}$ $(1; 2; -1)$, $\vec{e_2}$ $(3; 2; -3)$, $\vec{e_3}$ $(1; -2; -1)$;

2) $\vec{e_1}$ $(1; 3; 0)$, $\vec{e_2}$ $(2; -1; 1)$, $\vec{e_3}$ $(0; -1; 2)$.

Answer: 1) they do not form a basis; 2) $\vec{a} = 4\vec{e_1} + \vec{e_2} - \vec{e_3}$.

Problem 47. Show that the vectors $\vec{e_1}$ $(1; 0; 0)$, $\vec{e_2}$ $(1; 1; 0)$, $\vec{e_3}$ $(1; 1; 1)$ form a basis in space. Write the expansion of the vector $\vec{a} = -2\vec{i} - \vec{k}$ in this basis.

Answer: $\vec{a} = -2\vec{e_1} + \vec{e_2} - \vec{e_3}$.

Problem 48. Find the distance between the points *A* and *B* whose coordinates are given:

1) *A* $(-1; 2)$, *B* $(5; 10)$; 2) *A* $(1; 2)$, *B* $(2; 2)$;
3) *A* $(4; -2; 3)$, *B* $(4; 5; 2)$; 4) *A* $(3; -3; -7)$, *B* $(1; -4; -5)$.

***Guideline*:** Find the length of the vector $\vec{AB}$.

Answer: 1) 10; 2) 1; 3) $\sqrt{50}$; 4) 3.

1.7 SCALAR PRODUCT OF VECTORS

Definition. Product of the lengths of two vectors and the cosine of the angle between them is said to be a scalar product of these vectors. We will denote scalar product of the vectors $\vec{a}$ and $\vec{b}$ by $\left(\vec{a},\vec{b}\right)$. By definition

$$\left(\vec{a},\vec{b}\right)=\left|\vec{a}\right|\cdot\left|\vec{b}\right|\cos\phi .$$

Scalar product is a number. Scalar product has the following features:

1) $\left(\vec{a},\vec{b}\right)=\left(\vec{b},\vec{a}\right)$;

2) $\left(\lambda\vec{a},\vec{b}\right)=\left(\vec{a},\lambda\vec{b}\right)=\lambda\left(\vec{a},\vec{b}\right)$, λ is a number;

3) $\left(\vec{a}+\vec{b},\vec{c}\right)=\left(\vec{a},\vec{c}\right)+\left(\vec{b},\vec{c}\right)$.

It is clear that for any $\vec{a}$ vector

$$\left(\vec{a},\vec{a}\right)=\left|\vec{a}\right|\cdot\left|\vec{a}\right|\cos 0^{\circ}=\left|\vec{a}\right|^{2}.$$

When $\vec{b}$ and $\vec{b}$ vectors are perpendicular, then

$$\left(\vec{a},\vec{b}\right)=\left|\vec{a}\right|\cdot\left|\vec{b}\right|\cos 90^{\circ}=0.$$

If $\vec{i},\ \vec{j},\ \vec{k}$ is an orthonormal basis, then

$$\left(\vec{i},\vec{i}\right)=\left(\vec{j},\vec{j}\right)=\left(\vec{k},\vec{k}\right)=1,\left(\vec{i},\vec{j}\right)=\left(\vec{i},\vec{k}\right)=\left(\vec{j},\vec{k}\right)=0 .$$

Scalar product of the vectors $\vec{a}(a_x, a_y, a_z)$ and $\vec{b}(b_x, b_y, b_z)$ are expressed by the coordinates of these vectors by means of the formula

$$\left(\vec{a}, \vec{b}\right) = a_x b_x + a_y b_y + a_z b_z .$$

The cosine of the angle between the vectors $\vec{a}$ and $\vec{b}$ with known coordinates may be found by the following formula

$$\cos\varphi = \frac{\left(\vec{a}, \vec{b}\right)}{\left|\vec{a}\right| \cdot \left|\vec{b}\right|} = \frac{a_x b_x + a_y b_y + a_z b_z}{\sqrt{a_x^2 + a_y^2 + a_z^2} \cdot \sqrt{b_x^2 + b_y^2 + b_z^2}} .$$

Problems to be solved in auditorium

Problem 49. Find the scalar product of the vectors $\vec{a}$ and $\vec{b}$:

1) $\left|\vec{a}\right| = 3,\ \left|\vec{b}\right| = 1,\qquad \left(\vec{a}\,\hat{,}\,\vec{b}\right) = 45^{\circ}$;

2) $\left|\vec{a}\right| = 6,\ \left|\vec{b}\right| = 7,\qquad \left(\vec{a}\,\hat{,}\,\vec{b}\right) = 120^{\circ}$.

Answer: 1) $\frac{3\sqrt{2}}{2}$; 2) -21.

Problem 50. Find the scalar product of the vectors $\vec{b}$ and $\vec{b}$:

1) $\vec{a} = 4\vec{i} - \vec{j},\ \ \vec{b} = -\vec{i} - 7\vec{j}$; 2) $\vec{a}(2;\ 1),\ \ \vec{b}(1;\ -3)$;

3) $\vec{a}(3;\ 2;\ -5),\ \ \vec{b}(10;\ 1;\ 2)$; 4) $\vec{a} = 2\vec{i} + \vec{j} + 5\vec{k},\ \ \vec{b} = 7\vec{i} - 9\vec{j} - \vec{k}$.

Answer: 1) 3; 2) −1; 3) 22; 4) 0.

Problem 51. Find the angle between the vectors $\vec{a}$ and $\vec{b}$:

1) $\vec{a}(1;\ 2),\ \vec{b}(2;\ 4)$; 2) $\vec{a}(1;\ 2),\ \vec{b}(-2;\ 1)$;

3) $\vec{a}(1;\ 0;\ 0), \vec{b}(1;\ 1;\ 0)$; 4) $\vec{a}(1;\ -1;\ 1), \vec{b}(4;\ 4;\ -4)$.

Answer: 1) 0°; 2) 90°; 3) 45°; 4) *arccos*(−1/3).

Problem 52. Knowing that the vectors $\vec{a}=\vec{e_1}+2\vec{e_2},\ \vec{b}=5\vec{e_1}-4\vec{e_2}$ are perpendicular, find the angle between the unit vectors $\vec{e_1}$ and $\vec{e_2}$.

Solution. By the condition, as $\left|\vec{e_1}\right|=\left|\vec{e_2}\right|=1$

$$\left(\vec{e_1},\vec{e_2}\right)=\left|\vec{e_1}\right|\cdot\left|\vec{e_2}\right|\cos\left(\widehat{\vec{e_1},\vec{e_2}}\right)=\cos\left(\widehat{\vec{e_1},\vec{e_2}}\right).$$

Thus, if product of the vectors $\vec{e_1}$ and $\vec{e_2}$ are known, we can find the angle between them.

By the condition and features of scalar product,

$$\left(\vec{a},\vec{b}\right)=\left(\vec{e_1}+2\vec{e_2},\ 5\vec{e_1}-4\vec{e_2}\right)=5\left(\vec{e_1},\vec{e_1}\right)-4\left(\vec{e_1},\vec{e_2}\right)+$$

$$+10\left(\vec{e_2},\vec{e_1}\right)-8\left(\vec{e_2},\vec{e_2}\right)=5-4\left(\vec{e_1},\vec{e_2}\right)+10\left(\vec{e_1},\vec{e_2}\right)-8=$$

$$=6\left(\vec{e_1},\vec{e_2}\right)-3\Rightarrow\left(\vec{a},\vec{b}\right)=6\left(\vec{e_1},\vec{e_2}\right)-3.$$

Ontheother,asthevectors $\vec{a}$ and $\vec{b}$ areperpendicular, $\left(\vec{a},\vec{b}\right)=0$. According to the obtained expression of, $\left(\vec{a},\vec{b}\right)$ we get $6\left(\vec{e_1},\vec{e_2}\right)-3=0\Rightarrow\left(\vec{e_1},\vec{e_2}\right)=\frac{1}{2}$.

Above we showed that $\left(\vec{e_1},\vec{e_2}\right)=\cos(\widehat{e_1,e_2})$.

Hence, we can find $\cos\left(\widehat{\vec{e_1},\vec{e_2}}\right)=\frac{1}{2}\Rightarrow\left(\widehat{\vec{e_1},\vec{e_2}}\right)=60^\circ$.

Problem 53. Knowing $\left|\vec{a}\right| = 3$, $\left|\vec{b}\right| = 1$, $\left|\vec{c}\right| = 4$, and $\vec{a} + \vec{b} + \vec{c} = \vec{0}$, find the sum of $\left(\vec{a}, \vec{b}\right) + \left(\vec{b}, \vec{c}\right) + \left(\vec{c}, \vec{a}\right)$.

Guideline: Multiplying sequentially the both hand sides of the equality by the vectors $\vec{a} + \vec{b} + \vec{c} = \vec{0}$, put together the obtained three equalities $\vec{a}, \vec{b}, \vec{c}$.

Answer: −13.

Problem 54. Find the coordinates of the vector $\vec{x}$ collinear to the vector $\vec{a}$ (2; 1; −1) and satisfying the condition $\left(\vec{a}, \vec{x}\right) = 3$.

Answer: $\vec{x}\left(1; \frac{1}{2}; -\frac{1}{2}\right)$.

Problem 55. The vector $\vec{x}$ is perpendicular to the vectors $\vec{a}_1$ (2; 3; −1), $\vec{a}_2$ (1; −2; 3), and satisfies the condition $\left(\vec{x}, 2\vec{i} - \vec{j} + \vec{k}\right) = -6$. Find the coordinates of $\vec{x}$.

Answer: $\vec{x}$ (−3; 3; 3).

Problem 56. Find the lengths and inner angles of triangle with vertices at the points A (−1, −2, 4), B (−4, −2, 0), C (3, −2, 1).

Answer: $\left|\vec{AB}\right| = 5$, $\left|\vec{BC}\right| = 5\sqrt{2}$, $\left|\vec{AC}\right| = 5$, $\hat{A} = \frac{\pi}{2}$, $\hat{B} = \hat{C} = \frac{\pi}{4}$.

Home tasks

Problem 57. Find the scalar product of the vectors $\vec{a}$ and $\vec{b}$:

1) $\vec{a} = 3\vec{i} - 2\vec{j}$, $\vec{b} = \vec{i} + \vec{j}$; 2) $a = 5\vec{i}$, $\vec{b} = -\vec{i} - 2\vec{j}$;

3) $\vec{a} = 2\vec{i} - \vec{k} + \vec{j}$, $\vec{b} = \vec{k} - 3\vec{j}$; 4) $\vec{a} = \vec{i} + 2\vec{j} + 3\vec{k}$, $\vec{b} = 2\vec{j} + \vec{k}$.

Answer: 1) 1; 2) –5; 3) –4; 4) 7.

Problem 58. Find the angle between the vectors $\vec{a} = -\vec{i} + \vec{j}$ and $\vec{b} = \vec{i} - 2\vec{j} + 2\vec{k}$·

Answer: 135°.

Problem 59. Find the inner angles of a triangle with vertices at the points *A* (2; −1; 3), *B* (1; 1; 1), *C* (0; 0; 5).

Answer: $\hat{A} = 90^{\circ}$, $\hat{B} = \hat{C} = 45^{\circ}$.

Problem 60. Find the angle between the diagonals of a parallelogram constructed on the vectors $\vec{a} = 2\vec{i} + \vec{j}$ and $\vec{b} = -2\vec{j} + \vec{k}$.

Answer: 90°.

Problem 61. When $\vec{e_1}$, $\vec{e_2}$ are unit vectors with angle of 60° between them, find the lengths of diagonals of parallelograms constructed on the vectors $\vec{a} = 2\vec{e_1} + \vec{e_2}$ and $\vec{b} = \vec{e_1} - 2\vec{e_2}$.

Guideline: For finding the lengths of the vectors $\vec{d_1} = \vec{a} + \vec{b} = 3\vec{e_1} - \vec{e_2}$, $\vec{d_2} = \vec{a} - \vec{b} = \vec{e_1} + 3\vec{e_2}$, use the formulas $\left|\vec{d_1}\right| = \sqrt{\left(\vec{d_1}, \vec{d_1}\right)}$, $\left|\vec{d_2}\right| = \sqrt{\left(\vec{d_2}, \vec{d_2}\right)}$.

Answer: $\sqrt{7}$, $\sqrt{13}$.

Problem 62. Knowing $\left|\vec{a_1}\right| = 3$, $\left|\vec{a_2}\right| = 5$, find the values of α at which the vectors $\vec{a_1} + \alpha\,\vec{a_2}$ and $\vec{a_1} - \alpha\,\vec{a_2}$ are perpendicular.

Answer: $\alpha = \pm\frac{3}{5}$.

1.8 VECTORIAL PRODUCT OF VECTORS

Definition. Looking from the end of the vector $\vec{c}$ among non-coplanar three $\vec{a}$, $\vec{b}$, $\vec{c}$ vectors with the origins at the same point, if the direction of the small angle for taking $\vec{a}$ to $\vec{b}$ is in the counter clockwise, it is said that $\vec{a}, \vec{b}, \vec{c}$ vectors form right orientation, in the contrary case the left orientation triple.

Definition. The vector $\vec{c}$ satisfying the following three conditions is said to be vectorial product of the vectors $\vec{a}$ and $\vec{b}$:

1) Numerically, the length of $\vec{c}$ is equal to the area of the parallelogram constructed on $\vec{a}$ and $\vec{b}$, that is,

$$\left|\vec{c}\right| = \left|\vec{a}\right| \cdot \left|\vec{b}\right| \sin\varphi \,;$$

2) The vector $\vec{c}$ is perpendicular to the plane where the parallelogram constructed on the vectors $\vec{a}$ and $\vec{b}$ is located;

3) $\vec{a}, \vec{b}, \vec{c}$ form a right orientation triple.

We will denote vectorial product of the vectors $\vec{a}$ and $\vec{b}$ by $\vec{a} \times \vec{b}$. Vectorial product has the following features:

1) $\vec{a} \times \vec{b} = -\vec{b} \times \vec{a}$;

2) $\left(\lambda \vec{a}\right) \times \vec{b} = \vec{a} \times \left(\lambda\, \vec{b}\right) = \lambda \cdot \left(\vec{a} \times \vec{b}\right)$, λ is a number;

3) $\vec{a} \times \left(\vec{b} + \vec{c}\right) = \vec{a} \times \vec{b} + \vec{a} \times \vec{c}$;

4) If $\vec{a} \times \vec{b} = \vec{0}$, either $\vec{a} = 0$, or $\vec{b} = 0$, or $\vec{a} \parallel \vec{b}$.

For vectorial products of orthonormal basis vectors $\vec{i}, \vec{j}, \vec{k}$ the following relations are valid:

$$\vec{i} \times \vec{i} = \vec{j} \times \vec{j} = \vec{k} \times \vec{k} = \vec{0},$$

$$\vec{i} \times \vec{j} = \vec{k}, \quad \vec{j} \times \vec{k} = \vec{i}, \quad \vec{k} \times \vec{i} = \vec{j}.$$

Vectorial product of the vectors

$$\vec{a} = a_x \vec{i} + a_y \vec{j} + a_z \vec{k}\ ,\ \vec{b} = b_x \vec{i} + b_y \vec{j} + b_z \vec{k}$$

with known coordinates may be found by the following formula

$$\vec{a} \times \vec{b} = \begin{vmatrix} \vec{i} & \vec{j} & \vec{k} \\ a_x & a_y & a_z \\ b_x & b_y & b_z \end{vmatrix}.$$

Problems to be solved in auditorium

Problem 63. Open the parentheses and simplify the expressions:

$$1)\ \vec{i} \times \left(\vec{j} + \vec{k}\right) + \vec{j} \times \left(\vec{i} + \vec{k}\right) + \vec{k} \times \left(\vec{i} + \vec{j}\right);$$

$$2)\ \vec{i} \times \left(\vec{j} - \vec{k}\right) + \vec{j} \times \left(\vec{k} - \vec{i}\right) + \vec{k} \times \left(\vec{i} - \vec{j}\right).$$

Answer: 1) $\vec{0}$; 2) $2\cdot\left(\vec{i} + \vec{j} + \vec{k}\right)$.

Problem 64. Find vectorial product of the vectors $\vec{a}$ and $\vec{b}$ with known coordinates:

1) $\vec{a} = 4\vec{i} + 9\vec{j} + \vec{k}$; $\vec{b} = -3\vec{i} - 2\vec{j} + 5\vec{k}$;

2) $\vec{a}\ (1; -5; -1)$, $\vec{b}\ (2; -3; 3)$;

3) $\vec{a}\ (2; 3; 0)$, $\vec{b}\ (0; 3; 2)$;

4) $\vec{a} = 3\vec{i}$; $\vec{b} = 2\vec{k}$; 5) $\vec{a}\ (1; 1; 0)$, $\vec{b}\ (1; -1; 0)$.

Answer: 1) $47\vec{i} - 23\vec{j} + 19\vec{k}$; 2) $-18\vec{i} - 5\vec{j} + 7\vec{k}$;
3) $6\vec{i} - 4\vec{j} + 6\vec{k}$; 4) $-6\vec{j}$; 5) $-2\vec{k}$.

Problem 65. Find the area of the parallelogram constructed on the vectors $\vec{a} = 2\vec{i} + 3\vec{j} + 5\vec{k}$ and $\vec{b} = \vec{i} + 2\vec{j} + \vec{k}$.

Answer: $\sqrt{59}$ square unit.

Problem 66. For $\left|\vec{a}\right| = \left|\vec{b}\right| = 5$ the vectors $\vec{a}$ and $\vec{b}$ form angle of 45°. Find the area of the triangle constructed on the vectors $\vec{a} - 2\vec{b}$ and $3\vec{a} + 2\vec{b}$.

Solution. The area S of the triangle constructed on the vectors $\vec{a} - 2\vec{b}$ and $3\vec{a} + 2\vec{b}$ equals the half of the area of the parallelogram constructed on these vectors. Numerically, the area of this parallelogram equals the length of the vector that is the vectorial product of the vectors $\vec{a} - 2\vec{b}$ and $3\vec{a} + 2\vec{b}$. So,

$$S = \frac{1}{2} \cdot \left| \left(\vec{a} - 2\vec{b} \right) \times \left(3\vec{a} + 2\vec{b} \right) \right|.$$

According to features of vectorial product and the problem condition, we get:

$$S = \frac{1}{2} \cdot \left| 3\left(\vec{a} \times \vec{a} \right) + 2\left(\vec{a} \times \vec{b} \right) - 6\left(\vec{b} \times \vec{a} \right) - 4\left(\vec{b} \times \vec{b} \right) \right| =$$

$$= \frac{1}{2} \cdot \left| 2\left(\vec{a} \times \vec{b} \right) + 6\left(\vec{a} \times \vec{b} \right) \right| = \frac{1}{2} \left| 8\left(\vec{a} \times \vec{b} \right) \right| = 4 \cdot \left| \vec{a} \times \vec{b} \right| =$$

$$= 4 \cdot \left| \vec{a} \right| \cdot \left| \vec{b} \right| \sin \left(\widehat{\vec{a}, \vec{b}} \right) = 4 \cdot 5 \cdot 5 \cdot \sin 45^{\circ} = 100 \cdot \frac{\sqrt{2}}{2} = 50\sqrt{2}.$$

Problem 67. Find the area of the triangle with the vertices at the points A (1; 1; 1), B (2; 3; 4), C (4; 3; 2).

Guideline: The area of a triangle may be found as $\frac{1}{2}\left| \overrightarrow{AB} \times \overrightarrow{AC} \right|$.

Answer: $2\sqrt{6}$ square unit.

Problem 68. For $\vec{a}(2; 1; -3)$, $\vec{b}(1; -1; 1)$ find the vectors

$$\vec{a} \times \left(\vec{a} + \vec{b}\right) + \vec{a} \times \left(\vec{a} \times \vec{b}\right).$$

Answer: $-20\vec{i} + 7\vec{j} - 11\vec{k}$.

Home tasks

Problem 69. Open the parenthesis and simplify the expression:

1) $\vec{i} \times \left(\vec{j} + \vec{k}\right) - \vec{j} \times \left(\vec{i} + \vec{k}\right) + \vec{k} \times \left(\vec{i} + \vec{j} + \vec{k}\right)$;

2) $\left(\vec{a} + \vec{b} + \vec{c}\right) \times \vec{c} + \left(\vec{a} + \vec{b} + \vec{c}\right) \times \vec{b} + \left(\vec{b} - \vec{c}\right) \times \vec{a}$;

3) $\left(2\vec{a} + \vec{b}\right) \times \left(\vec{c} - \vec{a}\right) + \left(\vec{b} + \vec{c}\right) \times \left(\vec{a} + \vec{b}\right)$.

Answer: 1) $2\left(\vec{k} - \vec{i}\right)$; 2) $2\left(\vec{a} \times \vec{c}\right)$; 3) $\vec{a} \times \vec{c}$.

Problem 70. Find vectorial products of the vectors $\vec{a}$ and $\vec{b}$ with known coordinates:

1) $\vec{a} = 3\vec{i} - 2\vec{j} + 4\vec{k}$, $\vec{b} = -2\vec{i} + \vec{j} - 3\vec{k}$;

2) $\vec{a} = -5\vec{i} + 5\vec{j} - \vec{k}$, $\vec{b} = -\vec{i} - 5\vec{j} + 5\vec{k}$;

3) $\vec{a} = 2\vec{k}$, $\vec{b} = 2\vec{j}$; 4) $\vec{a}(1; -1; 0)$, $\vec{b}(0; 1; -1)$;

5) $\vec{a}(1; 0; 1)$, $\vec{b}(0; 1; 0)$; 6) $\vec{a}(0; 1; 1)$, $\vec{b}(0; 0; 1)$.

Answer:
1) $2\vec{i} + \vec{j} - \vec{k}$; 2) $20\vec{i} + 26\vec{j} + 30\vec{k}$; 3) $4\vec{i}$;
4) $\vec{i} + \vec{j} + \vec{k}$; 5) $-\vec{i} + \vec{k}$; 6) $\vec{i}$.

Problem 71. The vectors $\vec{a}$ (3; −1; 2) and $\vec{b}$ (1; 2; −1) are given. Find the following vectorial products:

1) $\vec{a} \times \vec{b}$; 2) $\left(2\vec{a} + \vec{b}\right) \times \vec{b}$; 3) $\left(2\vec{a} - \vec{b}\right) \times \left(2\vec{a} + \vec{b}\right)$.

Answer:

$$1)\ -3\vec{i}+5\vec{j}+7\vec{k};\ \ 2)-6\vec{i}+10\vec{j}+14\vec{k};\ \ 3)-12\vec{i}+20\vec{j}+28\vec{k}.$$

Problem 72. Knowing $|\vec{a}|=4, |\vec{b}|=5, \left(\widehat{\vec{a}, \vec{b}}\right)=30^{\circ}$, find the area of the parallelogram constructed on the vectors $\vec{a}-2\vec{b}$ and $3\vec{a}+2\vec{b}$:

Answer: 80 square unit.

Problem 73. Find the area of a triangle with vertices at the points A (1; 2; 3), B (3; 2; 1), C(1; −1; 1).

Answer: $\sqrt{22}$ square unit.

Problem 74. For $\vec{a}=3\vec{i}-\vec{j}+\vec{k}$ and $\vec{b}=\vec{i}+2\vec{k}$, at which values of α and β, the vector $\alpha\vec{i}+3\vec{j}+\beta\vec{k}$ will be collinear to the vector $\vec{a}\times\vec{b}$?

Answer: $\alpha=-6, \beta=21$.

1.9 MIXED PRODUCT OF VECTORS

Definition. Scalar product of vectorial product of the vectors $\vec{a}$ and $\vec{b}$ by the vector $\vec{c}$ is said to be mixed product of the vectors $\vec{a}\,\vec{b}\,\vec{c}$.

The mixed product of the vectors $\vec{a}, \vec{b}, \vec{c}$ will be denoted as $\vec{a}\,\vec{b}\,\vec{c}$. By definition

$$\vec{a}\,\vec{b}\,\vec{c}=\left(\vec{a}\times\vec{b}, \vec{c}\right).$$

It is clear that mixed product of arbitrary three vectors will give a number. We can find the mixed product of the vectors

$$\vec{a}=a_x\vec{i}+a_y\vec{i}+a_z\vec{k},\ \vec{b}=b_x\vec{i}+b_y\vec{i}+b_z\vec{k},$$
$$\vec{c}=c_x\vec{i}+c_y\vec{i}+c_z\vec{k}$$

by the formula

$$\vec{a}\,\vec{b}\,\vec{c} = \begin{vmatrix} a_x & a_y & a_z \\ b_x & b_y & b_z \\ c_x & c_y & c_z \end{vmatrix}.$$

Mixed product has the following features:

1. When replacing the vectors in the mixed product circularly, the mixed product does not change, that is,

$$\vec{a}\,\vec{b}\,\vec{c} = \vec{b}\,\vec{c}\,\vec{a} = \vec{c}\,\vec{a}\,\vec{b};$$

2. Replacing any two vectors in mixed product, the mixed product changes only its sign, that is,

$$\vec{a}\,\vec{b}\,\vec{c} = -\vec{b}\,\vec{a}\,\vec{c},\ \vec{a}\,\vec{b}\,\vec{c} = -\vec{c}\,\vec{b}\,\vec{a},\ \vec{a}\,\vec{b}\,\vec{c} = -\vec{a}\,\vec{c}\,\vec{b}.$$

The modulus of three vectors of mixed product equals numerically to the volume of the parallelepiped constructed on these vectors.

A necessary and sufficient condition for three vectors $\vec{a}, \vec{b}, \vec{c}$ be coplanar is $\vec{a}\,\vec{b}\,\vec{c} = 0$.

Problems to be solved in auditorium

Problem 75. The vectors $\vec{a_1}, \vec{a_2}, \vec{a_3}$ are relatively perpendicular and form a right orientation triple $\left|\vec{a_1}\right| = 4,\ \left|\vec{a_2}\right| = 2,\ \left|\vec{a_3}\right| = 3$. Find mixed product of these vectors.

Solution. By definition of mixed product, $\vec{a_1}\,\vec{a_2}\,\vec{a_3} = \left(\vec{a_1} \times \vec{a_2}, \vec{a_3}\right)$.

Denote the vectorial product of $\vec{a_1}$ and $\vec{a_2}$ by $\vec{c}$ $\vec{c} = \vec{a_1} \times \vec{a_2}$. By definition of vectorial product and the condition,

$$\left|\vec{c}\right| = \left|\vec{a_1} \times \vec{a_2}\right| = \left|\vec{a_1}\right| \cdot \left|\vec{a_2}\right| \cdot \sin\left(\vec{a_1}, \stackrel{\wedge}{\ } \vec{a_2}\right) = 4 \cdot 2 \cdot \sin 90^{\circ} = 8.$$

The vector $\vec{c}$ is perpendicular to the plane where the vectors $\vec{a}_1$ and $\vec{a}_2$ are arranged, and the vectors $\vec{a}_1, \vec{a}_2, \vec{c}$ form a right orientation triple. Hence and from the condition of the problem we get that $\vec{c}$ and $\vec{a}_3$ have the same direction. Therefore, $\left(\vec{c}, \vec{a}_3\right) = 0^\circ$.

Hence, by definition of scalar product and the condition of the problem we get:

$$\left(\vec{c}, \vec{a}_3\right) = \left|\vec{A}\right| \cdot \left|\vec{a}_3\right| \cdot \cos\left(\vec{c}, \vec{a}_3\right) = 8 \cdot 3 \cdot \cos 0^\circ = 24.$$

Thus, $\vec{a}_1 \vec{a}_2 \vec{a}_3 = \left(\vec{a}_1 \times \vec{a}_2, \vec{a}_3\right) = \left(\vec{c}, \vec{a}_3\right) = 24.$

Problem 76. Find mixed product of the vectors $\vec{a}, \vec{b}, \vec{c}$ with the given coordinates:

1) $\vec{a} = \vec{i} + 2\vec{j} + 3\vec{k},\ \vec{b} = -\vec{i} - \vec{j} + 2\vec{k},\ \vec{c} = 3\vec{i} + \vec{j} - 2\vec{k}$;

2) $\vec{a} = 2\vec{i} - 3\vec{j} + \vec{k},\ \vec{b} = 5\vec{i} - \vec{j} + 4\vec{k},\ \vec{c} = -\vec{i} + \vec{j} - 3\vec{k}$;

3) $\vec{a}(1;\ 2;\ 3),\ \vec{b}(0;\ 4;\ 5),\ \vec{c}(0;\ 0;\ 6)$;

4) $\vec{a} = 2\vec{k},\ \vec{b} = 2\vec{j},\ \vec{c} = 2\vec{i}$.

Answer: 1) 14; 2) –21; 3) 24; 4) –8.

Problem 77. Find the volume of a triangular pyramid with the vertices at the points $A\ (1;\ 1;\ 1)$, $B\ (3;\ 2;\ 1)$, $C\ (5;\ 3;\ 2)$, $D\ (3;\ 4;\ 5)$.

Guideline: Take into account that the volume of the sought-for parallelepiped equals $\frac{1}{6}$ of the parallelepiped constructed on the vectors $\vec{AB}, \vec{AC}, \vec{AD}$.

Answer: $\frac{2}{3}$.

Problem 78. Determine if the vectors $\vec{a}$ (–1;1;–1), $\vec{b}$ (–1;1;–1), $\vec{c}$ (1;1;1) are coplanar.

Answer: They are noncoplanar.

Problem 79. Prove that the points A (5; 7; −2), B (3; 1; −1), C (9; 4; −4), D (1; 5; 0) are arranged on the same plane.

Solution. When the points A, B, C, and D are on the same plane, the vectors $\vec{AB}(-2;-6;1)$, $\vec{AC}(4;-3;-2)$, $\vec{AD}(-4;-2;2)$ are coplanar. When these vectors are coplanar, their mixed product should equal zero. Let us calculate this mixed product:

$$\vec{AB}\cdot\vec{AC}\cdot\vec{AD}=\begin{vmatrix}-2 & -6 & 1\\ 4 & -3 & -2\\ -4 & -2 & 2\end{vmatrix}.$$

If we multiply the third column of this determinant by –2, we get the first column, that is, the first and third columns are proportional. Therefore, the determinant equals zero. So $\vec{AB}\cdot\vec{AC}\cdot\vec{AD}=0$. This means that the points A, B, C, and D, are on the same plane.

Home tasks

Problem 80. Find the mixed product of the vectors $\vec{a}$, $\vec{b}$, $\vec{c}$ with known coordinates:

1) $\vec{a}=\vec{i}+2\vec{j}+3\vec{k}$; $\vec{b}=4\vec{i}+5\vec{j}+6\vec{k}$; $\vec{c}=7\vec{i}+8\vec{j}+9\vec{k}$;

2) $\vec{a}(3;4;-5)$, $\vec{b}(8;7;-2)$, $\vec{c}(2;-1;8)$;

3) $\vec{a}=\vec{i}+\vec{k}$, $\vec{b}=-\vec{i}-\vec{j}$, $\vec{c}=-\vec{j}-\vec{k}$;

4) $\vec{a}=2\vec{i}-\vec{j}$; $\vec{b}=2\vec{j}-\vec{k}$; $\vec{c}=2\vec{k}-\vec{i}$.

Answer: 1) 0; 2) 0; 3) 2; 4) 7.

Problem 81. Verify if the vectors $\vec{a}, \vec{b}, \vec{c}$ are coplanar:

1) $\vec{a}=\vec{i}+\vec{j}+\vec{k},\ \vec{b}=-2\vec{i}+3\vec{j}-2\vec{k},\ \vec{c}=3\vec{i}-2\vec{j}+3\vec{k};$
2) $\vec{a}=2\vec{i}-\vec{j}+3\vec{k},\ \vec{b}=\vec{i}+2\vec{j}-3\vec{k},\ \vec{c}=\vec{i}+3\vec{j}-2\vec{k}\cdot$

Answer: 1) are coplanar; 2) are noncoplanar.

Problem 82. Determine if the points A, B, C, and D with known coordinates are on the same plane:

1) $A\,(1;0;1), B\,(2;1;-2), C\,(1;2;0), D\,(-1;1;-1);$
2) $A\,(1;2;-1), B\,(0;1;5), C\,(-1;2;1), D\,(2;1;3).$

Answer: 1) are not on the same plane;
2) are on the same plane.

Problem 83. Find the volume of a triangular pyramid with vertices at the points A (2; 0; 0), B (0; 3; 0), C (0; 0; 6), D (2; 3; 8).

Answer: 14 cubic unit.

Problem 84. The vectors $\vec{a}, \vec{b}, \vec{c}$ are given, and $\left(\widehat{\vec{a},\vec{b}}\right)=\left(\widehat{\vec{c},\vec{a}\times\vec{b}}\right)=\alpha$.
Prove that $\vec{a}\,\vec{b}\,\vec{c}=\frac{1}{2}\sin 2\alpha.$

Guideline: Use the definition of mixed product $\left(\vec{a}\,\vec{b}\,\vec{c}=\left(\vec{a}\times\vec{b},\vec{c}\right)\right)$.

Problem 85. The vectors $\vec{a}, \vec{b}, \vec{c}$ are noncoplanar. At which values of λ the vectors

$$\vec{a}+2\vec{b}+\lambda\vec{c},\ 4\vec{a}+5\vec{b}+6\vec{c},\ 7\vec{a}+8\vec{b}+\lambda^2\vec{c}$$

may be coplanar.

Guideline: take the vectors $\vec{a}, \vec{b}, \vec{c}$ as a basis.

Answer: $\lambda = 3$ and $\lambda = -4$.

1.10 STRAIGHT LINE EQUATIONS ON A PLANE

$y = kx + b$ is said to be angular coefficient equation of a straight line. Here, k is called an angular coefficient and equals the tangent of the angle a formed by this straight line and positive direction of the abscissa axis: α: $k = tg\alpha$ and b indicates the ordinate of the point that intersections the ordinate axis of this straight line.

The equation of the straight line with angular coefficient k and passing through the point M_0 (x_0, y_0) is in the form

$$y - y_0 = k\,(x - x_0). \tag{1.12}$$

The tangent of the angle φ between the straight lines l_1 and l_2 given by angular coefficient equations $y = k_1\, x + b_1$ and $y = k_2\, x + b_2$ may be found by the formula

$$tg\varphi = \left|\frac{k_1 - k_2}{1 + k_1 k_2}\right| . \tag{1.13}$$

For $l_1 \,||\, l_2$, $k_1 = k_2$.

For $l_1 \perp l_2$, $1 + k_1 k_2 = 0 \Rightarrow k_2 = -\dfrac{1}{k_1}$.

The equation of the straight line passing through the points $M_1(x_1, y_1)$ and $M_2(x_2, y_2)$ is in the form

$$\frac{y - y_1}{y_2 - y_1} = \frac{x - x_1}{x_2 - x_1}. \tag{1.14}$$

The equation

$$Ax + By + C = 0,\ (A^2 + B^2 \neq 0) \tag{1.15}$$

is called a general equation of a straight line. The vector $\vec{n}\,(A;B)$ is said to be a normal vector of this straight line.

The equation of a straight line passing through the point M_0 (x_0, y_0) and possessing the normal vector $\vec{n}\,(A;B)$ is written as follows

$$A(x - x_0) + B(y - y_0) = 0. \tag{1.16}$$

The equation of a straight line passing through the point M_0 (x_0, y_0) and parallel to the vector $\vec{a}$ $(m; n)$ is in the form

$$\frac{x-x_0}{m}=\frac{y-y_0}{n}. \tag{1.17}$$

The last equality is called a canonical equation of the straight line.

Parametric equations of a straight line are written in the form

$$\begin{cases} x = x_0 + mt, \\ y = y_0 + nt, \qquad t \in (-\infty, +\infty) \end{cases}. \tag{1.18}$$

The straight line determined by this equation also passes through the point M_0 (x_0, y_0) and is parallel to the vector $\vec{a}$ $(m; n)$.

If in the general equation (4) of straight line $A^2 + B^2 = 1$ and $C<0$, (1.15) is said to be a normal equation of the straight line. If $A^2 + B^2 = 1$ and $C>0$, we can multiply all the terms of the equation by -1 and reduce it to the normal form. For example, in the equation $\frac{3}{5}x-\frac{4}{5}y+6=0$, $A=\frac{3}{5}$; $B=-\frac{4}{5}$; $A^2+B^2=\frac{9}{25}+\frac{16}{25}=1$, but as $C=5>0$, this is not a normal equation. The equation $-\frac{3}{5}x+\frac{4}{5}y-6=0$ obtained by multiplying all terms by (-1) is a normal equation.

In eq 1.15, we take $A^2 + B_2 =1$, $C<0$ for $A = \cos \alpha$, $B = \sin \alpha$, $C = -p$ $(p > 0)$ and write the equation of the straight line in the form

$$x \cos \alpha + y = \sin \alpha - p = 0. \tag{1.19}$$

Here α is an angle formed by the perpendicular drawn to the given straight line from the origin of coordinates and the positive direction of the abscissa axis, p is the distance from the origin of coordinates to this straight line.

For $C < 0$, $A^2 + B^2 \neq 1$, for reducing equation (1.15) to such a form, all terms of eq 1.15 should be divided into $\sqrt{A^2+B^2}$:

$$\frac{A}{\sqrt{A^2+B^2}}x+\frac{B}{\sqrt{A^2+B^2}}y+\frac{C}{\sqrt{A^2+B^2}}=0.$$

The equation of the straight line intersecting the abscissa axis at the point whose abscissa is a, the ordinate axis whose ordinate is b, is written in the form

$$\frac{x}{a}+\frac{y}{b}=1. \tag{1.20}$$

Equation 1.20 is said to be a piecewise equation of the straight line.

The distance from the point M_0 (x_0, y_0) to the straight line given by normal eq 1.17 is found by the formula

$$d=|x_0 \cos\alpha + y_0 \sin\alpha - p| , \tag{1.21}$$

the distance from this point to the straight line given by common eq 1.15 is found by the formula

$$d=\frac{|Ax_0 + By_0 + C|}{\sqrt{A^2+B^2}} . \tag{1.22}$$

The equation of a straight line given in certain kind may be transformed to its equation in another kind. The way for obtaining normal equation from general equation was given above.

If a straight line is given by the angular coefficient equation $y = kx + b$, its general equation will be in the form $kx - y + b = 0$. For example, the equation of straight line with angular coefficient $y = 3x - 2$ may be written in the general form as $3x - y - 2 = 0$.

In order to transform general equation to angular coefficient eq 1.15 should be solved with respect to y:

$$Ax+By+C=0 \Rightarrow By=-Ax-C \Rightarrow y=-\frac{A}{B}x-\frac{C}{A};\ (A \neq 0)\cdot$$

For example, if we transform the equation $3x - 2y + 6 = 0$ to an angular coefficient equation, we get

$$3x-2y+6=0 \Rightarrow 2y=3x+6 \Rightarrow y=\frac{3}{2}x+3\cdot$$

Let us transform general eq 1.15 to piecewise equation:

$$Ax+By+C=0 \Rightarrow Ax+By=-C \Rightarrow -\frac{A}{C}x-\frac{B}{C}y=1 \Rightarrow$$

$$\Rightarrow \frac{x}{-\frac{C}{A}}+\frac{y}{-\frac{C}{B}}=1,\ (A \neq 0, B \neq 0, C \neq 0).$$

For example, let us write the equation $2x - 3y + 2 = 0$ in the form of a piecewise equation:

$$2x-3y+2=0 \Rightarrow 2x-3y=-2 \Rightarrow -x+\frac{3}{2}y=1 \Rightarrow \frac{x}{-1}+\frac{y}{\frac{2}{3}}=1.$$

For writing piecewise equation in the general form, we must multiply its both sides by ab:

$$\frac{x}{a}+\frac{y}{b}=1 \Rightarrow bx+ay=ab \Rightarrow bx+ay-ab=0.$$

If the straight lines l_1 and l_2 are given by the general equations $A_1 x + B_1 y + C_1 = 0$, $A_2 x + B_2 y + C_2 = 0$, respectively, the angle φ between them may be found as an angle between the normal $\vec{n_1}(A_1, B_1)$, $\vec{n_2}(A_2, B_2)$ of these straight lines by the formula

$$\cos\varphi = \frac{\left(\vec{n_1}, \vec{n_2}\right)}{\left|\vec{n_1}\right| \cdot \left|\vec{n_2}\right|} = \frac{A_1 A_2 + B_1 B_2}{\sqrt{A_1^2 + B_1^2} \cdot \sqrt{A_2^2 + B_2^2}}. \tag{1.23}$$

For $l_1 \perp l_2$, from eq 1.23 we get the perpendicularity condition of two straight lines given by general equations:

$$\varphi = \frac{\pi}{2} \Rightarrow \cos\varphi = 0 \Rightarrow A_1 A_2 + B_1 B_2 = 0. \tag{1.24}$$

For $l_1 \,||\, l_2$, the vectors $\vec{n_1}$ and $\vec{n_2}$ are collinear. Hence we get that the condition of parallelecity of two straight lines given by general equations is in the form

$$\frac{A_1}{A_2} = \frac{B_1}{B_2}. \tag{1.25}$$

Problems to be solved in auditorium

Problem 86. Write the equation of a straight line intersecting the OY axis at the point $(0;-3)$ and forming an angle of 45° with positive direction of the axis OX.

Answer: $y = x - 3$.

Problem 87. Write the equation of straight line intersecting the axis OY at the point (0;2) and forming an angle of 30° with the positive direction of the axis OY.

Answer: $y=\sqrt{3}x+2$.

Problem 88. Write the equation of a straight line passing through the point $A(2;3)$ and forming angle of 135° with the positive direction of the abscissa axis.

Answer: $y = -x + 5$.

Problem 89. Write the equation of a straight line that passes through the point $A(-3;4)$ and is parallel to the straight line $x - 2y + 5 = 0$.

Answer: $y=\frac{1}{2}x+5\frac{1}{2}$.

Problem 90. Find the angles between angular coefficient equations and straight lines:

1) $y = -x + 3$ and $y = -x + 2$; 2) $y = 2x - 7$ and $y=-\frac{1}{2}x+9$;

3) $y = -3x + 5$ and $y = -3x + 3$; 4) $y = -0{,}4x - 1$ and $y=2\frac{1}{2}x+1$;

5) $y = -3x - 9$ and $y=\frac{1}{2}x+2$; 6) $y = 2x + 3$ and $y=\frac{2-\sqrt{3}}{1+2\sqrt{3}}x-5$.

Answer: 1) 0°; 2) 90°; 3) 0°; 4) 90°; 5) 45°; 6) 60°.

Guideline: Use the parallelecity condition of straight lines given by angular coefficient equation in 1) and 3), the perpendicularity conditions in eqs 1.2 and 1.4. Use formula (1.13) in eqs 1.5 and 1.6.

Problem 91. Write the equation of the straight line passing through the given two points with known coordinates:

1) A(3; −1), B(−2; 5); 2) A(1; 2), B(−3; −4).

Answer: 1) $6x + 5y - 13 = 0$; 2) $3x - 2y + 1 = 0$.

Guideline: Use formula (1.14).

Problem 92. The general form of the straight line in the form $12x-5y65=0$ is given. Write the angular coefficient, piecewise and normal equations of this straight line.

Answer: $y=\frac{12}{5}x-13, \frac{x}{\frac{65}{12}}+\frac{y}{-13}=1; \frac{12}{13}x-\frac{5}{13}y-5=0.$

Problem 93. Write the equation of the straight line passing through the point $A(-8;9)$ and separating triangle with 6 square unit area from the coordinate.

Answer: $3x+4y-12=0$, $27x+16y+72=0$.

Problem 94. Write the equation of a straight line passing through the point $M_0(x_0, y_0)$ and possessing the normal vector $\vec{n}(A,B)$:

$$1)\, M_0(2;1),\ \vec{n}(2;5);\ 2)\, M_0(-1;2),\ \vec{n}(2;3).$$

Answer: 1) $2x+5y-9=0$; 2) $2x+3y-4=0$.

Guideline: Use formula (1.16).

Problem 95. Write the equation of a straight line passing through the point $M_0(x_0, y_0)$ and parallel to the vector $\vec{a}(m;n)$

$$1)\, M_0(-1;2),\ \vec{a}(3;4);\ 2)\, M_0(3;1),\ \vec{a}(-1;-2).$$

Answer: 1) $4x-3y+10=0$; 2) $2x-y-5=0$.

Guideline: Use (1.17).

Problem 96. Write the parametric equations of a straight line passing through the point $M_0(0;-1)$ and parallel to the vector $\vec{a}(2;-6)$. Show that the point $M(-1;2)$ is on this straight line.

Answer: $x=2t$, $y=-1-6t$.

Guideline: Use (1.18).

Problem 97. The perpendicular drawn from the origin of coordinates to the straight line forms angle of 30° with positive direction of the abscissa axis. Distance from the origin of coordinates to this straight line is 5 unit. Write the equation of this straight line.

Answer: $\sqrt{3}x+y-10=0$.

Guideline: Use (8).

Problem 98. Find the distance from the point M_0 (–2;3) to the straight line given by the equation $3x + 4y - 1 = 0$.

Answer: 1.

Guideline: Use formula (1.22).

Problem 99. Find the angle between the straight lines $\sqrt{3}x+y-2=0$ and $\sqrt{3}x-y+2=0$.

Answer: 60°.

Guideline: Use formula (1.23).

Problem 100. Determine mutual states of straight lines given by general equations:

1) $2x - 3y + 5 = 0$ and $-4x + 6y - 1 = 0$;
2) $2x - y - 4 = 0$ and $x + 2y - 3 = 0$;
3) $3x + 5y - 7 = 0$ and $-6x + 3{,}6y - 0{,}2 = 0$;
4) $x - y - 1 = 0$ and $-2x + 2y - 1 = 0$.

Answer: 1) they are parallel; 2) they are perpendicular; 3) they are perpendicular; 4) they are parallel.

Guideline: Verify conditions (1.24) and (1.25).

Problem 101. Find the length of the height of a triangle with vertices at the point *A(–3;0), B(2;5), C(3;2)* drawn from the vertex *B* to the side *AC*.

Answer: $\sqrt{10}$

Problem 102. The equation $\left(\sqrt{3}+1\right)x-\left(\sqrt{3}-1\right)y- -5+\sqrt{3}=0$, the equation $y - 3 = 0$ of the side *AC*, and the coordinates *D(5; 6)* of the foot of the height *AB* and *AD* of a triangle is given. Find inner angles of this triangle.

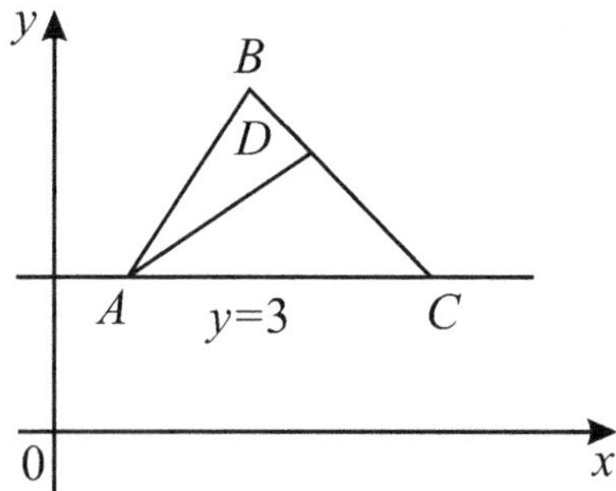

Solution.

It is clear that by finding the angles between the straight lines *AD* and *AB*, *AC* we can calculate all inner angles of the triangle. For that, it suffices to know the equation of *AD*.

We solve jointly the equations of the straight lines *AB* and *AC* and find the coordinates of *A* that is the intersection of these straight lines:

$$\begin{cases} y = 3, \\ \left(\sqrt{3}+1\right)x-\left(\sqrt{3}-1\right)y-5+\sqrt{3}=0 \end{cases} \Rightarrow$$

$$\Rightarrow\left(\sqrt{3}+1\right)x-\left(\sqrt{3}-1\right)\cdot 3-5+\sqrt{3}=0\Rightarrow$$

$$\Rightarrow\left(\sqrt{3}+1\right)x=2\left(\sqrt{3}+1\right)\Rightarrow x=2. \text{ So, } A(2;3).$$

Now write the equation of the straight line *AD*:

$$\frac{x-2}{5-2}=\frac{y-3}{6-3}\Rightarrow\frac{x-2}{3}=\frac{y-3}{3}\Rightarrow x-2=y-3\Rightarrow y=x+1.$$

So, the equation of *AD* is in the form $y = x + 1$.

It is seen from the equation $y - 3 = 0$ of the straight line *AC* that this straight line is parallel to the abscissa axis. Therefore, the angle between *AD* and *AC* equals the angle between *AD* and the abscissa axis. It is seen from the equation of *AD* that the angle formed by it and the abscissa axis is 45° ($K_{AD} = 1 \Rightarrow tg\varphi = 1$).

So, $\left(A\hat{D}, AC\right)=45^\circ$. From the right triangle *ACD* we get, $\hat{C}=90^\circ-45^\circ=45^\circ$.

Now find the angle between AB and AD. We write the equation of AB in the form

$$y = \frac{\sqrt{3}+1}{\sqrt{3}-1}x - \frac{5-\sqrt{3}}{\sqrt{3}-1}$$

and find its angular coefficient:

$$K_{AB} = \frac{\sqrt{3}+1}{\sqrt{3}-1}.$$

As above noted, $K_{AB} = 1$. By formula (1.13)

$$tg\left(A\hat{B},\, AD\right) = \frac{K_{AB} - K_{AD}}{1 + K_{AB}\cdot K_{AD}} = \frac{\dfrac{\sqrt{3}+1}{\sqrt{3}-1} - 1}{1 + 1\cdot\dfrac{\sqrt{3}+1}{\sqrt{3}-1}} =$$

$$= \frac{\sqrt{3}+1-\sqrt{3}+1}{\sqrt{3}-1+\sqrt{3}+1} = \frac{2}{2\sqrt{3}} = \frac{1}{\sqrt{3}}.$$

Hence we get $\left(A\hat{B},\, AD\right) = 30^\circ$.

The values of the angles A and B may be found as follows:

$$\hat{B} = 90^\circ - \left(A\hat{B},\, AD\right) = 90^\circ - 30^\circ = 60^\circ,$$

$$\hat{A} = \left(A\hat{D},\, AC\right) + \left(A\hat{D},\, AB\right) = 45^\circ + 30^\circ = 75^\circ.$$

Answer: 75°, 60°, 45°.

Home tasks

Problem 103. Write the equation of the straight line passing through the point $A(-3;4)$ and perpendicular to the straight line $x - 2y + 5 = 0$.

Answer: $y = -2x - 2$.

Problem 104. At which values of a, the straight lines given by the equations ax – $4y = 6$ or $x - xy = 3$:

1) intersect; 2) are parallel; 3) coincide.

Answer: 1) for $a \neq \pm 2$; 2) for $a = -2$; 3) for $a = 2$.

Guideline: study the system formed by the equations of these straight lines.

Problem 105. Find the angles between the straight lines given by the following equations:

1) $y = 2x + 7$; and $y = 2x + 7$; 2) $y = 5x + 9$ and $y = -0,\ 2x - 3$;

3) $y = 2x + 1$ and $y = \frac{1}{3}x - 10$; 4) $3x - 5y - 3 =$ and $-x + \frac{5}{3}y - 3 = 0$;

5) $9x - 2x - 7 = 0$ and $-2x + 9y - 5 = 0$;

6) $10x - 2y - 1 = 0$ and $2x - 3y - 4 = 0$.

Answer: 1) 0°; 2) 90° 3) 45°; 4) 0°; 5) 90°; 6) 45°.

Problem 106. Write the equation of the straight line passing through the point $A(3;1)$ and forming angle of 45° with the straight line $3x = y + 2$.

Answer: $2x + y - 7 = 0$, $x - 2y - 1 = 0$.

Guideline: Look for the equation of a straight line in the form $y - 1 = K_1 (x - 3)$, take $\varphi = 45°$, $K_2 = 3$, find K_1 by formula (1.13).

Problem 107. Find the acute angle formed by the straight line passing through the points $A(2;\sqrt{3})$ $B(3;2\sqrt{3})$ and the ordinate axis.

Answer: 30°.

Problem 108. Write the equation of the straight line passing through the point $A(3;4)$ and separating a triangle of 6 square unit from the coordinate quarter.

Answer: $\frac{x}{-3} + \frac{y}{2} = 1,\ \frac{x}{1,5} + \frac{y}{-4} = 1.$

Problem 109. Write the equation of a straight line passing through the point M_0 (x_0, y_0) and possessing the normal vector $\vec{n}(A, B)$. Find the distance from the point $A(a, b)$ to this straight line:

$1)\, M_0(1;-1),\ \vec{n}(3;4),\ A(2;-3);$

$2)\, M_0(-2;1),\ \vec{n}(6;8),\ A(-1;1).$

Answer: 1)1; 2) 0,6. ***Guideline:*** Use formulas (1.16) and (1.22).

Problem 110. Find the distance from the point $M(1;1)$ to the straight line given by the parametric equations $x = -1 + 2t$, $y = 2 + t$.

Answer: $\frac{4}{\sqrt{5}}$. ***Guideline:*** Write the general solution of the given straight line and use formula (1.22).

Problem 111. A straight line is given by parametric equations $x = 1 - t$, $y = 2 - t$. Find the angle formed by the abscissa axis of the perpendicular drawn to this straight line and the distance from the origin of coordinates to this straight line.

Answer: 135°, $\frac{1}{\sqrt{2}}$. ***Guideline:*** Write the normal equation of the straight line given by parametric equations.

Problem 112. The length of diagonal of the rhomb on the abscissa axis with intersection of diagonals at the origin of coordinates equals 6, its area equals 12 square unit. Write the equations of the sides of this rhomb.

Answer: $2x+3y-6=0,\ 2x-3y+6=0,$
$2x+3y+6=0,\ 2x-3y-6=0.$

Problem 113. Prove that a triangle on straight lines with the sides given by the equations $x+\sqrt{3}y+1=0$, $\sqrt{3}x+y+1=0$, $x-y-10=0$ is equilateral, and find its vertex angles.

Answer: 30°.

Problem 114. Find the length of the height drawn to the hypotenuse of a right triangle with vertices at the points $A(2;5)$, $B(5;1)$, $C(5;5)$.

Answer: 2,4.

1.11 PLANE AND STRAIGHT LINE EQUATIONS IN SPACE

The equation of the plane passing through the point $M_0(x_0, y_0, z_0)$ and parallel to noncollinear vectors $\vec{a}(a_1, a_2, a_3)$ and $\vec{b}(b_1, b_2, b_3)$ may be written in the form

$$\begin{vmatrix} x-x_0 & y-y_0 & z-z_0 \\ a_1 & a_2 & a_3 \\ b_1 & b_2 & b_3 \end{vmatrix} = 0. \tag{1.26}$$

Here, x, y, z denote the coordinates of the arbitrary point M on the plane α. As $x_0, y_0, z_0, a_1, a_2, a_3, b_1, b_2, b_3$ are the given numbers, we can open the third-order determinant at the left-hand side of (1.26), cancel the same terms and get

$$Ax + By + Cz + D = 0. \tag{1.27}$$

Here A, B, C, D are the known numbers.

Equation (1.27) is called a general equation of the plane.

If we denote $\overrightarrow{OM_0} = \vec{r}_0(x_0, y_0, z_0)$, $\overrightarrow{OM} = \vec{r}(x, y, z)$, expand the vector $\vec{r} - \vec{r}_0$ in noncollinear vectors $\vec{a}, \vec{b}$, we can write

$$\vec{r} - \vec{r}_0 = u\vec{a} + v\vec{b}. \tag{1.28}$$

Here u, v are certain parameters.

Equation (1.28) is called a parametric equation of the plane α in the vector form.

If we express eq. (1.28) by coordinates, we get

$$\begin{cases} x - x_0 = a_1 u + b_1 v, \\ y - y_0 = a_2 u + b_2 v, \\ z - z_0 = a_3 u + b_3 v. \end{cases} \tag{1.29}$$

Equation (1.29) is called parametric equations of the plane α.

When the plane α is given by the general equation (1.27), for writing its parametric equations we should take $x = u$, $y = v$ and solve eq. (1.27) with respect to z. Then parametric equations of α will be written in the form

$$\begin{cases} x = u, \\ y = v, \\ z = -\dfrac{A}{C}x - \dfrac{B}{C}y - \dfrac{D}{C}, \end{cases} \qquad (C \neq 0).$$

When a plane is given by parametric eq. (1.29), from the first and second equations of formula (1.29) we find the expressions of u and v by x,y, write them in the third equation, and get the general equation of this plane.

For example, write general equation of the plane given by the parametric equation

$$\begin{cases} x = 1 + u - v, \\ y = 2 + u + 2v, \\ z = -1 - u + 2v. \end{cases}$$

For that at first from the system

$$\begin{cases} x = 1 + u - v, \\ y = 2 + u + 2v \end{cases} \Rightarrow \begin{cases} u - v = x - 1, \\ u + 2v = y - 2 \end{cases}$$

we denote u and v by x,y:

$$u = \frac{1}{3}(2x + y - 4), \quad v = \frac{1}{3}(y - x - 1) \cdot$$

Take into account these expressions of u and v in the third one of the given parametric equations:

$$z = -1 - u + 2v \Rightarrow z = -1 - \frac{1}{3} \cdot (2x + y - 4) + 2 \cdot \frac{1}{3} \cdot (y - x - 1) \Rightarrow$$
$$\Rightarrow 3z = -3 - 2x - y + 4 + 2y - 2x - 2 \Rightarrow 4x - y + 3z + 1 = 0.$$

The last equality is the general equation of the plane given by parametric equations.

The vector $\vec{n} = A\vec{i} + B\vec{j} + C\vec{k}$ whose coordinates are the numbers A, B, C in general eq. (1.27) of the plane is called a normal vector of this plane. The vector $\vec{n}$ is perpendicular to the plane.

For $\left|\vec{n}\right| = 1$, that is, $A^2 + B^2 + C^2 = 1$ and $D < 0$, (1.27) is called a normal equation of the plane. For $\left|\vec{n}\right| = 1$, $A = \cos\alpha$, $B = \cos\beta$, and $C = \cos\gamma$. Here α, β, γ denote the angles formed by the vector $\vec{n}$ and the axes, respectively. Taking into account these expressions of A, B, C in (1.27), taking $D = -p$, $(p > 0)$, we can write the normal equation of the plane in the form

$$x\cos\alpha + y\cos\beta + z\cos\gamma - p = 0. \tag{1.30}$$

Here p equals the distance from the origin of coordinates to this plane.

For $A^2 + B^2 + C^2 \neq 1$, by multiplying all terms of eq. (1.27) by the number

$$\mu = \pm\frac{1}{\left|\vec{n}\right|} = \pm\frac{1}{\sqrt{A^2 + B^2 + C^2}},$$

we can transform the general equation of the plane to a normal equation. The sign of μ in eq. (1.27) should be taken opposite to the sign of D in eq. (1.27).

The equation of a plane passing through the point $M_0(x_0, y_0, z_0)$ and perpendicular to the vector $\vec{n}(A, B, C)$ is in the form

$$A(x - x_0) + B(y - y_0) + C(z - z_0) = 0. \tag{1.31}$$

The equation of the plane intersecting the axis OX at the point with abscissa α, the axis OY with ordinate b, the axis OZ with applicate c is written as

$$\frac{x}{a} + \frac{y}{b} + \frac{z}{c} = 1. \tag{1.32}$$

(1.32) is called a piecewise equation of the plane.

When $D \neq 0$, for transforming general eq. (1.27) to piecewise equation, we should divide all terms of eq. (1.27) into D:

$$Ax + By + Cz + D = 0 \Rightarrow -\frac{A}{D}x - \frac{B}{D}y - \frac{C}{D}z = 1 \Rightarrow$$

$$\Rightarrow \frac{x}{-\frac{D}{A}} + \frac{y}{-\frac{D}{B}} + \frac{z}{-\frac{D}{C}} = 1.$$

For transforming the piecewise equation of plane (1.32) into a general equation, we should multiply all terms of eq. (1.32) by abc ($abc \neq 0$):

$$\frac{x}{a}+\frac{y}{b}+\frac{z}{c}=1 \Rightarrow bcx+acy+abz-abc=0.$$

The angle φ given by the equations $A_1x + B_1y + C_1z + D_1 = 0$ and $A_2x + B_2y + C_2z + D_2 = 0$ may be found as an angle between their normal vectors $\vec{n_1}(A_1, B_1, C_1)$, $\vec{n_2}(A_2, B_2, C_2)$ from the formula

$$\cos\varphi=\frac{\left(\vec{n_1},\vec{n_2}\right)}{\left|\vec{n_1}\right|\cdot\left|\vec{n_2}\right|}=\frac{A_1A_2+B_1B_2+C_1C_2}{\sqrt{A_1^2+B_1^2+C_1^2}\cdot\sqrt{A_2^2+B_2^2+C_2^2}}. \tag{1.33}$$

When these planes are parallel, as their normal vectors are collinear, the parallelism condition of the planes given by the above general equations is in the form

$$\frac{A_1}{A_2}=\frac{B_1}{B_2}=\frac{C_1}{C_2}. \tag{1.34}$$

Condition of perpendicularity of planes is obtained from relation (1.33) for $\varphi = 90°$:

$$A_1A_2 + B_1B_2 + C_1C_2 = 0. \tag{1.35}$$

The distance from the arbitrary point M_0 (x_0, y_0, z_0) in space to the plane given by eq. (1.27) may be found by the formula

$$d=\frac{\left|Ax_0+By_0+Cz_0+D\right|}{\sqrt{A^2+B^2+C^2}}. \tag{1.36}$$

The equation of the plane passing through the three points $M_1(x_1,y_1,z_1)$, $M_2(x_2,y_2,z_2)$, $M_3(x_3,y_3,z_3)$ not arranged on a straight line in space is written as follows

$$\begin{vmatrix} x-x_1 & y-y_1 & z-z_1 \\ x_2-x_1 & y_2-y_1 & z_2-z_1 \\ x_3-x_1 & y_3-y_1 & z_2-z_1 \end{vmatrix}=0. \tag{1.37}$$

When the numbers A_1, B_1, C_1 are not proportional to the numbers A_2, B_2, C_2, respectively, the system of linear equation consisting of the equations of two planes

$$\begin{cases} A_1x+B_1y+C_1z+D_1=0, \\ A_2x+B_2y+C_2z+D_2=0 \end{cases} \tag{1.38}$$

has a finite number of solutions and the points set whose coordinates satisfy (1.38) in space, is a straight line that is the intersection of these planes. Therefore, when the coefficients satisfy the above condition, (1.38) is said to be general conditions of a straight line in space.

The vectorial equation of a straight plane that passes through the points M_0 (x_0, y_0, z_0) and is parallel to the vector $\vec{a}(a_1, a_2, a_3)$ is written as follows

$$\vec{r} = \vec{r_0} + \vec{a}\,t. \tag{1.39}$$

Here $\vec{r_0} = \overrightarrow{OM_0}, \vec{r} = \overrightarrow{OM}, M(x, y, z)$ is an arbitrary point on this straight line.

Express (1.39) by coordinates:

$$\begin{cases} x = x_0 + a_1t, \\ y = y_0 + a_2t, \\ z = z_0 + a_3t. \end{cases} \tag{1.40}$$

Equation (1.40) is called parametric equations of a straight line. $\vec{a}$ is said to be direction vector of this straight line.

Having found t from each equation of (1.40), equating the obtained expressions, we get

$$\frac{x-x_0}{a_1} = \frac{y-y_0}{a_2} = \frac{z-z_0}{a_3}. \tag{1.41}$$

(1.41) is said to be a canonical equation of a straight line.

The equation of a straight line passing through two different points M_1 (x_1, y_1, z_1) and M_2 (x_2, y_2, z_2) in space is written in the form

$$\frac{x-x_1}{x_2-x_1} = \frac{y-y_1}{y_2-y_1} = \frac{z-z_1}{z_2-z_1}. \tag{1.42}$$

If these two different points have equal ones among its coordinates, (1.42) is not most used. For example, if $x_1 = x_2$, $y_1 \neq y_2$, $z_1 \neq z_2$, then the equation of a straight line passing through the points $M_1\ (x_1, y_1, z_1)$ and $M_2\ (x_1, y_2, z_2)$ is written as follows

$$x = x_1,\ \frac{y - y_1}{y_2 - y_1} = \frac{z - z_1}{z_2 - z_1}.$$

In the case when other coordinates are equal, the equations may be represented by the similar rule.

The cosines of the angles α, β, γ by eqs (1.40) or (1.41) in the space and the coordinate axes of the given straight line may be found by the relations

$$\cos\alpha = \frac{a_1}{\sqrt{a_1^2 + a_2^2 + a_3^2}},\ \cos\beta = \frac{a_2}{\sqrt{a_1^2 + a_2^2 + a_3^2}},$$
$$\cos\gamma = \frac{a_3}{\sqrt{a_1^2 + a_2^2 + a_3^2}}, \tag{1.43}$$

respectively.

If the straight line l_1 passing through the point $M_1\ (x_1, y_1, z_1)$ is given by the canonical $\frac{x - x_1}{a_1} = \frac{y - y_1}{a_2} = \frac{z - z_1}{a_3}$ or parametric equation $\begin{cases} x = x_1 + a_1 t, \\ y = y_1 + a_2 t, \\ z = z_1 + a_3 t \end{cases}$ the straight line l_2 passing through the point $M_2\ (x_2, y_2, z_2)$ is given by the canonic $\frac{x - x_2}{b_1} = \frac{y - y_2}{b_2} = \frac{z - z_2}{b_3}$ or parametric equations, the angle φ between the straight lines l_1 and l_2 is calculated as an angle between the direction vectors $\vec{a}(a_1, a_2, a_3)$, $\vec{b}(b_1, b_2, b_3)$ of these straight lines by the formula

$$\cos\phi = \frac{\left(\vec{a}, \vec{b}\right)}{\left|\vec{a}\right| \cdot \left|\vec{b}\right|} = \frac{a_1 b_1 + a_2 b_2 + a_3 b_3}{\sqrt{a_1^2 + a_2^2 + a_3^2} \cdot \sqrt{b_1^2 + b_2^2 + b_3^2}}. \tag{1.44}$$

When the straight lines l_1 and l_2 are parallel, their direction vectors are collinear, the condition of parallelism of these two straight lines in the space is written as

$$\frac{a_1}{b_1} = \frac{a_2}{b_2} = \frac{a_3}{b_3}. \tag{1.45}$$

When l_1 and l_2 are mutually perpendicular ($\varphi = 90° \Rightarrow \cos\varphi = 0$), from (1.44) we get that the perpendicularity condition of two straight lines in the space is in the form

$$a_1b_1 + a_2b_2 + a_3b_3 = 0. \tag{1.46}$$

Mutual state of the straight lines l_1 and l_2 given in space by the above equations may be studied with the help of the following two matrices made of the coordinates of the vectors $\vec{a}, \vec{b}$ and

$$\overrightarrow{M_1M_2}(x_2 - x_1, y_2 - y_1, z_2 - z_1).$$

Denote:

$$M_1 = \begin{pmatrix} a_1 & a_2 & a_3 \\ b_1 & b_2 & b_3 \end{pmatrix}, \quad M_2 = \begin{pmatrix} a_1 & a_2 & a_3 \\ b_1 & b_2 & b_3 \\ x_2 - x_1 & y_2 - y_1 & z_2 - z_1 \end{pmatrix}.$$

rank $M_1 = r_1$, *rank* $M_2 = r_2$. The following cases are possible:

1. $r_2 = 3$. In this case by definition of rank

$$\Delta = \begin{vmatrix} a_1 & a_2 & a_3 \\ b_1 & b_2 & b_3 \\ x_2 - x_1 & y_2 - y_1 & z_2 - z_1 \end{vmatrix} \neq 0.$$

   This means that the vectors $\vec{a}, \vec{b}$, $\overrightarrow{M_1M_2}$ are noncoplanar. So l_1 and l_2 are crossing.

2. $r_2 = 2$, $r_1 = 2$. As $\Delta = 0$ the vectors $\vec{a}, \vec{b}$, $\overrightarrow{M_1M_2}$ are coplanar. So, the straight lines l_1 and l_2 are on the same plane. But it follows from

$r_1 = 2$ that $\vec{a}$ and $\vec{b}$ are not collinear. Therefore, in this case l_1 and l_2 intersect.

3. $r_2 = 2$, $r_1 = 1$. In this case, l_1 and l_2 are parallel.
4. $r_1 = r_2 = 1$. In this case l_1 and l_2 coincide.

A necessary and sufficient condition for arrangement of straight lines l_1 and l_2 on the same plane is satisfaction of the equality

$$\begin{vmatrix} a_1 & a_2 & a_3 \\ b_1 & b_2 & b_3 \\ x_2 - x_1 & y_2 - y_1 & z_2 - z_1 \end{vmatrix} = 0. \tag{1.47}$$

As the relation $\varphi = 90^0 - \left(\widehat{\vec{n}, \vec{a}}\right)$ is valid for the angle φ between the plane given by the equation $Ax + By + Cz + D = 0$ and the straight line $\frac{x - x_0}{a_1} = \frac{y - y_0}{a_2} = \frac{z - z_0}{a_3}$, this angle may be found by the formula

$$\sin\varphi = \sin\left(90^\circ - \left(\widehat{\vec{n}, \vec{a}}\right)\right) = \cos\left(\widehat{\vec{n}, \vec{a}}\right) =$$

$$= \frac{Aa_1 + Ba_2 + Ca_3}{\sqrt{A^2 + B^2 + C^2} \cdot \sqrt{a_1^2 + a_2^2 + a_3^2}}. \tag{1.48}$$

Here $\vec{n}(A, B, C)$ is a normal vector of the plane, $\vec{a}(a_1, a_2, a_3)$ is a direction vector of the straight line.

When a straight line is perpendicular to the plane, as $\vec{n}$ and $\vec{a}$ are collinear, the condition of perpendicularity of the straight line to the plane is obtained as follows

$$\frac{A}{a_1} = \frac{B}{a_2} = \frac{C}{a_3}. \tag{1.49}$$

When a straight line is parallel to the plane, as $\varphi = 0^\circ \Rightarrow \sin\varphi = 0$, the condition of parallelism of the straight line to a plane is obtained as

$$Aa_1 + Ba_2 + Ca_3 = 0. \tag{1.50}$$

For finding the coordinates of intersection points (if they exist) of a straight line and plane, their equations should be solved jointly. This time, it is convenient to use general equations of a plane and parametric equations of a straight line. If a plane is given by $Ax + By + Cz + D = 0$ a straight line by the equations $x = x_0 + a_1t$, $y = y_0 + a_2t$, $z = z_0 + a_3t$, then

1. for $Aa_1 + Ba_2 + Ca_3 \neq 0$, the straight line intersects the plane;
2. if the conditions $Aa_1 + Ba_2 + Ca_3 = 0$, and $Ax_0 + By_0 + Cz_3 D \neq 0$ are satisfied simultaneously, the straight line has no joint point with the plane and the straight line is parallel to the plane;
3. if both of the relations $Aa_1 + Ba_2 + Ca_3 = 0$, and $Ax_0 + By_0 + Cz_0$ $D = 0$ are satisfied, the straight line is on the plane.

Problems to be solved in auditorium

Problem 115. Write the equation of a plane that passes through the point M_0 and is parallel to noncollinear vectors $\vec{a}$ and $\vec{b}$:

1) $M_0\,(1;2;3)$, $\vec{a}\,(3;4;5)$, $\vec{b}\,(-2;-1;-3)$;
2) $M_0\,(-1;0;2)$, $\vec{a}\,(2;-1;1)$, $\vec{b}\,(3;1;-1)$.

Solution of 2): Take $x_0 = -1$, $y_0 = 0$, $z_0 = 2$, $a_1 = 2$; $a_2 = -1$, $a_3 = 1$, $b_1 = 3$, $b_2 = 1$, $b_3 = -1$ and use eq. (1.26):

$$\begin{vmatrix} x+1 & y & z-2 \\ 2 & -1 & 1 \\ 3 & 1 & -1 \end{vmatrix} = 0 \Rightarrow$$

$$\Rightarrow x+1+3y+2(z-2)+3(z-2)+2y-(x+1)=0 \Rightarrow$$

$$\Rightarrow 5y+5(z-2)=0 \Rightarrow y+z-2=0.$$

The last equation is the equation of the sought-for plane (verify if this plane passes through the point M_0 and is parallel to the vectors $\vec{a}$ and $\vec{b}$).

Answer: 1) $7x + y - 5z + 6 = 0$.

Problem 116. Write the equation of a plane that passes through the point $M_0(1;2;0)$ and is perpendicular to the vector $\vec{n}(-1;2;3)$.

Answer: $x - 2y - 3z + 3 = 0$. ***Guideline:*** Use (1.31).

Problem 117. Write the equation of a plane that passes through the points M_1 and M_2 with known coordinates and is parallel to the vector $\vec{a}$:

1) $M_1(1;2;0)$, $M_2(2;1;1)$, $\vec{a}\,(3;0;1)$;
2) $M_1(1;1;1)$, $M_2(2;3;-1)$, $\vec{a}\,(0;-1;2)$.

Solution of 1): As only one plane passes through two intersecting straight lines, a unique plane passes through noncollinear arbitrary two vectors. Therefore, as the vectors $\overrightarrow{M_1M_2}\,(1;-1;1)$ and $\vec{a}(3;0;1)$ are noncollinear, the problem has a unique solution.

According to definition of vectorial product, the vector $\vec{n} = \overrightarrow{M_1M_2} \times \vec{a}$ is perpendicular to the sought-for plane. According to the rule of finding vectorial product of two vectors with known coordinates,

$$\vec{n} = \overrightarrow{M_1M_2} \times \vec{a} = \begin{vmatrix} \vec{i} & \vec{j} & \vec{k} \\ 1 & -1 & 1 \\ 3 & 0 & 1 \end{vmatrix} = -\vec{i} + 3\vec{j} + 3\vec{k} - \vec{j} = -\vec{i} + 2\vec{j} + 3\vec{k}.$$

So, the sought-for plane is perpendicular to the vector $\vec{n_2}\,(2k;0;-k)$. If we take into account that this plane passes through the point $M_1(1;2;0)$ as well, we get that its equation is in the form

$$-1(x-1) + 2(y-2) + 3z = 0 \Rightarrow x - 2y - 3z + 3 = 0$$

(see problem 116).

Answer: 2) $2x - 2y - z + 1 = 0$.

Problem 118. Write the equation of a plane that passes through the point $A(1;-1;2)$ and is parallel to the plane given by the following equation:

$$1)\ x - 3y + 2z + 1 = 0;\ 2)\ 2x - z + 1 = 0.$$

Solution of 2): Normal vectors of two parallel planes must be collinear. As the normal vector of the given plane $2x - z + 1 = 0$ is $\vec{n_1}(2;0;-1)$, the coordinates of the normal vector $\vec{n_2}(A;B;C)$ of the plane parallel to this plane are in the form

$$A = 2k,\ B = 0\cdot k,\ C = -1\cdot k = -k,\ (k \neq 0).$$

Write the equation of a plane that passes through the point $A(1;-1;2)$ and has the normal vector $\vec{n_2}(2k;0;-k)$ (see eq. (1.31)):

$$2k(x-1) - k(z-2) = 0 \Rightarrow 2(x-1) - (z-2) = 0 \Rightarrow$$
$$\Rightarrow 2x - 2 - z + 2 = 0 \Rightarrow 2x - z = 0.$$

The last equality is the equation of the sought-for plane.

Answer:

Problem 119. Write the equation of a plane that passes through the point M_0 and is perpendicular to the given two planes:

1) $M_0(-1;-1;2), x - 2y + z - 4 = 0, x + 2y - 2z + 4 = 0;$

2) $M_0(0;0;5), x - y - z = 0, x - 2y = 0.$

Solution of 1): write the equation of a plane that passes through the point $M_0(-1;-1;2)$ and possesses the normal vector $\vec{n}(A;B;C)$ (see eq. (1.31)):

$$A(x+1) + B(y+1) + C(z-2) = 0. \qquad (1.51)$$

By the condition, this plane should be perpendicular to the planes given by the equations $x - 2y + z - 4 = 0$ and $x + 2y - 2z + 4 = 0$. From the condition of perpendicularity of two planes (see eq. (1.35)) we get:

$$\begin{cases} A\cdot 1 + B\cdot(-2) + C\cdot 1 = 0, \\ A\cdot 1 + B\cdot 2 + C\cdot(-2) = 0 \end{cases} \Rightarrow \begin{cases} A - 2B + C = 0, \\ A + 2B - 2C = 0. \end{cases}$$

From the last system denote A and B by C:

$$\begin{cases} 2A - C = 0, \\ 4B - 3C = 0 \end{cases} \Rightarrow \begin{cases} A = \dfrac{1}{2}C, \\ B = \dfrac{3}{4}C. \end{cases}$$

Take into account these expressions of A and B in eq. (1.51):

$$\frac{1}{2}C(x+1)+\frac{3}{4}C(y+1)+C(z-2)=0\Rightarrow$$
$$\Rightarrow 2(x+1)+3(y+1)+4(z-2)=0\Rightarrow 2x+3y+4z-3=0.$$

The last equality is the equation of the sought-for plane.

Answer: 2) $2x + y + z - 5 = 0$.

Problem 120. Write the equation of a plane that passes through the points M_1 and M_2 and is perpendicular to the given plane:

$$1)\ M_1(-1;-2;0),\ M_2(1;1;2),\ x+2y+2z-4=0;$$
$$2)\ M_1(0;1;2),\ M_2(-1;0;1),\ y-2z+7=0.$$

Solution of 1): Denote the normal vector of the sought-for plane by $\vec{n}$. As this plane passes both through the points M_1 (–1; –2;0), and M_2 (1;1;2), by eq. (1.6) we can write:

$$A(x+1)+B(y+2)+Cz=0\Rightarrow Ax+By+Cz+(A+2B)=0,$$
$$A(x-1)+B(y-1)+C(z-2)=0\Rightarrow Ax+By+Cz+(-A-B-2C)=0.$$

As the both of the last relations are the equations of the same plane, we compare free terms and get

$$A+2B=-A-B-2C\Rightarrow 2A+3B+2C=0. \qquad (1.52)$$

By the condition, the sought-for plane is perpendicular to the plane $x+2y+2z-4=0$ and according to the condition of perpendicularity of two planes (1.35) we can write

$$A+2B+2C=0\cdot \qquad (1.53)$$

In equalities (1.52) and (1.53) denote A and B by C:

$$\begin{cases}2A+3B+2C=0,\\ A+2B+2C=0\end{cases}\Rightarrow\begin{cases}A+B=0\\ A+2B+2C=0\end{cases}\Rightarrow\begin{cases}B=-A,\\ A-2A+2C=0\end{cases}\Rightarrow$$
$$\Rightarrow\begin{cases}B=-A,\\ A=2C\end{cases}\Rightarrow\begin{cases}A=2C,\\ B=-2C.\end{cases}$$

Write these expressions of A and B in one of the above equations of the sought-for plane:

$$Ax + By + Cz + (A + 2B) = 0 \Rightarrow 2Cx - 2Cy + Cz + 2C - 4C = 0 \Rightarrow$$
$$\Rightarrow 2x - 2y + z - 2 = 0.$$

The last equality is the equation of the sought-for plane.

Answer: 2) $3x - 2y - z + 4 = 0$.

Problem 121. Find the distance from the point $A(3;1;-1)$ to the plane given by the following equation:

$$1)\ x - y - 5z + 2 = 0;\quad 2)\ x - 2y + 2z + 7 = 0.$$

Answer: 1) $\sqrt{3}$; 2) 2. ***Guideline:*** Use formula (1.36).

Problem 122. Find the angle between the following equations and given planes:

1) $x + 4y - z + 1 = 0$ and $x + y - z - 3 = 0$;
2) $x + 2y - 2z = 0$ and $z = 5$;
3) $x + 3y - z + 1 = 0$ and $\begin{cases} x = 1 - u, \\ y = 2 - 3u - v, \\ z = 7 + u + v. \end{cases}$

Answer: 1) $arc\ \cos\left(\dfrac{\sqrt{6}}{3}\right)$; 2) $arc\ \cos\dfrac{2}{3}$; 3) $90°$.

Guideline: In 1) and 2) use direct formula (1.8), in 3) use it after writing the general equation of the second plane.

Problem 123. Find the distance between the parallel planes with given equations: 1) $6x - 3y + 2z + 5 = 0$, $6x - 3y + 2z - 9 = 0$; 2) $3x + 4z + 1 = 0$, $6x + 8z - 1 = 0$.

Solution of 1): We can find the distance between two parallel planes as a distance from arbitrary point on one of these planes to another plane. For writing coordinates of any point on a plane, we can give arbitrary values to any two coordinates and find the third coordinate from the equation of a plane.

If in the equation $6x - 3y + 2z + 5 = 0$ we take $x = 0, y = 0$, we get $z = -2.5$. So, the point $M_0(0;0;-2{,}5)$ is on the plane $6x - 3y + 2z + 5 = 0$. Calculate the distance from this point to the plane $6x - 3y + 2z - 9 = 0$ by formula (1.36):

$$d = \frac{|6 \cdot 0 - 3 \cdot 0 + 2 \cdot (-2{,}5) - 9|}{\sqrt{6^2 + (-3)^2 + 2^2}} = \frac{14}{7} = 2.$$

So, the distance between the given parallel planes is 2.

Answer: 2) $\frac{3}{10}$.

Problem 124. Write the equation of a plane that passes through the point $M(1;-3;5)$ and the segments separated by it from the axis OY and OZ are twice greater than separated from the axes OX.

Answer: $\frac{x}{2}+\frac{y}{4}+\frac{z}{4}=1.$

Guideline: Use the piecewise eq. (1.32) of the plane.

Problem 125. Write the equation of a straight line that passes through the points M_1 and M_2:

1) $M_1(1;3;-1)$, $M_2(4;2;1)$; 2) $M_1(3;2;5)$, $M_2(4;1;5)$;
3) $M_1(-1;1;2)$, $M_2(5;1;2)$; 4) $M_1(2;-1;3)$, $M_2(3;1;4)$.

Answer: 1) $\frac{x-1}{3}=\frac{y-3}{-1}=\frac{z+1}{2}$; 2) $\frac{x-3}{1}=\frac{y-2}{-1}$, $z=5$;

3) $y=1$, $z=2$; 4) $\frac{x-2}{1}=\frac{y+1}{2}=\frac{z-3}{1}$.

Guideline: In 1) and 4) use eq. (1.42).

Problem 126. Determine if three points are on one straight line. If they are not on one straight line, write the equation of a plane that passes through these three points:

1) $A(2;1;3)$, $B(-1;2;5)$, $C(3;0;1)$;
2) $A(1;-1;3)$, $B(2;3;4)$, $C(-1;1;2)$;
3) $A(1;1;2)$, $B(2;3;3)$, $C(-1;-3;0)$.

Answer:

1) they are not on one straight line, $2y-z+1=0$;
2) they are not on one straight line, $6x+y-10z+25=0$;
3) they are on one straight line.

Guideline: In order to verify if the given three points are on one straight line, we write the equation of the straight line that passes through any of these two points and verifies if the coordinates of the third point satisfy this equation.

Problem 127. Write canonical equations of a straight line given by parametric equation and general solutions of planes whose intersection line gives this straight line:

1) $x = 2+3t, y = 3-t, z = 1+t;$ 2) $x = 1+2t, y = 2+3t, z = 3-4t.$

Solution of 1): at first write the canonical equation of a straight line. For that from each of the given parametric equations of a straight line we find t and equate the expressions obtained for t:

$$t = \frac{x-2}{3},\ t = -(y-3),\ t = z-1 \Rightarrow$$

$$\Rightarrow \frac{x-2}{3} = -(y-3) = z-1 \Rightarrow \frac{x-2}{3} = \frac{y-3}{-1} = \frac{z-1}{1}.$$

The last relation is the canonical equation of the given straight line. From the canonical equation we get:

$$\frac{x-2}{3} = \frac{y-3}{-1} = \frac{z-1}{1} \Rightarrow \begin{cases} \dfrac{x-2}{3} = \dfrac{y-3}{-1}, \\ \dfrac{y-3}{-1} = \dfrac{z-1}{1} \end{cases} \Rightarrow$$

$$\Rightarrow \begin{cases} -x+2 = 3y-9, \\ y-3 = -z+1 \end{cases} \Rightarrow \begin{cases} x+3y-11 = 0, \\ y+z-4 = 0. \end{cases}$$

The last two equations are the general equations of planes whose intersection is the given straight line.

Answer: 2) $\begin{cases} 3x-2y+1 = 0, \\ 4y+3z-17 = 0. \end{cases}$

Problem 128. Write parametric and canonical equations of a straight line that is the intersection of planes given by general equations:

$$1)\begin{cases} x-y+2z+4 = 0, \\ -2x+y+z+3 = 0; \end{cases} \quad 2)\begin{cases} x+y+z+2 = 0, \\ 2x-2y-z-1 = 0. \end{cases}$$

Solution of 1): Take $z = t$ and denote x and y by t:

$$\begin{cases} x-y = -4-2z, \\ -2x+y = -3-z \end{cases} \Rightarrow \begin{cases} x-y = -4-2t, \\ -2x+y = -3-t \end{cases} \Rightarrow \begin{cases} -x = -7-3t, \\ y = x+4+2t \end{cases} \Rightarrow$$

$$\Rightarrow \begin{cases} x = 7+3t, \\ y = 7+3t+4+2t \end{cases} \Rightarrow \begin{cases} x = 7+3t, \\ y = 11+5t. \end{cases}$$

So, parametric equations of the given straight are in the form $x = 7 + 3t$, $y = 1 + 5t$, $z = t$.

From each of these three relations we find t, equate the obtained expressions and get the canonical equation:

$$t = \frac{x-7}{3}, t = \frac{y-11}{5}, t = z \Rightarrow \frac{x-7}{3} = \frac{y-11}{5} = \frac{z}{1}.$$

Answer: 2) $x = -\frac{3}{4} - \frac{1}{4}t,\ \ y = -\frac{5}{4} - \frac{3}{4}t,\ \ z = t;$

$$\frac{x+\frac{3}{4}}{-1} = \frac{y+\frac{5}{4}}{-3} = \frac{z}{4}.$$

Problem 129. Write the canonical equation of a straight line that passes through the point $M_0(2;0;-3)$ and is parallel

1) to the vector $a(2;-3;5)$;
2) to the straight line $\frac{x-1}{5} = \frac{y+2}{2} = \frac{z+1}{-1}$;
3) to the straight line $\begin{cases} 3x - y + 2z - 7 = 0, \\ x + 3y - 2z - 3 = 0 \end{cases}$;
4) to the straight line $x = -2 + t, y = 2t, z = 1 - \frac{1}{2}t$.

Answer: 1) $\frac{x-2}{2} = \frac{y}{-3} = \frac{z+3}{5}$; 2) $\frac{x-2}{5} = \frac{y}{2} = \frac{z+3}{-1}$;

3) $\frac{x-2}{-4} = \frac{y}{8} = \frac{z+3}{10}$; 4) $\frac{x-2}{1} = \frac{y}{2} = \frac{z+3}{-\frac{1}{2}}$.

Problem 130. Determine how the straight line given by the following equation is arranged with respect to the plane $x - 3y + z + 1 = 0$:

1) $\frac{x-1}{5} = \frac{y-1}{4} = \frac{z-1}{7}$;
2) $x = 2 + 3t, y = 7 + t, z = 1 + t$;
3) $x = 2, y = 5 + t, z = 4 + 3t$.

Solution of 2): It is clear that the coefficients in the plane equation $x - 3y + z + 1 = 0$ are $A = 1$, $B = -3$, $C = 1$, the coordinates of the direction vector $\vec{a}$ of the straight line $y = 7 + t, z = 1 + t, x = 2 + 3t$, are $a_1 = 3$, $a_2 = 1$, $a_3 = 1$.

As $Aa_1 + Ba_2 + Ca_3 = 1 \cdot 3 + (-3) \cdot 1 + 1 \cdot 1 = 1 \neq 0$, this straight line intersects the plane.

For finding the coordinates of the intersection point, we write the expressions of the variables x, y, z in the equation of the plane, and the expressions are given by t in the parametric equations of the straight line and solve the obtained equation with respect to t:

$$x - 3y + z + 1 = 0 \Rightarrow 2 + 3t - 3 \cdot (7 + t) + 1 + t = 0 \Rightarrow t_0 = 17.$$

Having substituted the found values of t_0 in the parametric equations of the straight line, we can determine the coordinates x_0, y_0, z_0 of the intersection points of the straight line and plane:

$$x_0 = 2 + 3t_0 = 2 + 3 \cdot 17 = 53, \quad y_0 = 7 + t_0 = 7 + 17 = 24,$$
$$z_0 = 1 + t_0 = 1 + 17 = 18.$$

So, straight lines intersect the plane at the point $M_0(53;24;18)$.

Answer: 1) the straight line is on the plane;
3) the straight line is parallel to the plane.

Problem 131. Determine mutual states of straight lines given by the following equations in space:

$$1) \begin{cases} x + z - 1 = 0, \\ 3x + y - z + 13 = 0 \end{cases} \text{and} \begin{cases} x - 2y + 3 = 0, \\ y + 2z - 8 = 0; \end{cases}$$

$$2)\; x = 3 + t,\; y = -1 + 2t,\; z = 4 \text{ and } \begin{cases} x + y - z = 0, \\ 2x - y + 2z = 0; \end{cases}$$

$$3)\; x = 2 + 4t, y = -6t, z = -1 - 8t \text{ and } x = 7 - 6t, y = 2 + 9t, z = 12t;$$

$$4)\; x = 9t, y = 5t, z = -3 + t \text{ and } \begin{cases} 2x - 3y - 3z - 9 = 0, \\ x - 2y + z + 3 = 0. \end{cases}$$

When the straight lines intersect, find the coordinates of the intersection point and the equation of the plane where these straight lines are arranged.

Solution of 1): At first we find parametric equations of both straight lines:

$$\begin{cases} x = 1 - z, \\ y = -13 - 3x + z, \\ z = t \end{cases} \Rightarrow \begin{cases} x = 1 - t, \\ y = -13 - 3(1-t) + t, \\ z = t \end{cases} \Rightarrow \begin{cases} x = 1 - t, \\ y = -16 + 4t, \\ z = t, \end{cases}$$

$$\begin{cases} x = -3 + 2y, \\ y = 8 - 2z, \\ z = t \end{cases} \Rightarrow \begin{cases} x = -3 + 2(8 - 2t), \\ y = 8 - 2t, \\ z = t \end{cases} \Rightarrow \begin{cases} x = 13 - 4t, \\ y = 8 - 2t, \\ z = t. \end{cases}$$

It is seen from the obtained equations that in this example,

$\vec{a}(-1;4;1)$, $\vec{b}(-4;-2;1)$, $M_1(1;-16;0)$, $M_2(13;8;0)$,

$\overrightarrow{M_1M_2}(12;24;0)$.

Verify if the vectors $\vec{a}$, $\vec{b}$, $\overrightarrow{M_1M_2}$ are coplanar:

$$\Delta = \begin{vmatrix} -1 & 4 & 1 \\ -4 & -2 & 1 \\ 12 & 24 & 0 \end{vmatrix} = 48 - 96 + 24 + 24 = 0.$$

So, $r_2 < 3$, that is, the vectors $\vec{a}$, $\vec{b}$, $\overrightarrow{M_1M_2}$ are coplanar.

Now verify if the vectors $\vec{a}$ and $\vec{b}$ are collinear.

$$M_1 = \begin{pmatrix} -1 & 4 & 1 \\ -4 & -2 & 1 \end{pmatrix} \to \begin{pmatrix} -1 & 4 & 1 \\ 0 & -18 & -3 \end{pmatrix} \Rightarrow rank\, M_1 = 2$$

So in this example, as $r_1 = 2, r_2 = 2$ the given straight lines intersect. For finding the coordinates of the intersection point we must solve jointly the given common or found parametric equations of these straight lines. For example, if we equate the expressions of x from parametric equations of both straight lines, we get $1 - t = 13 - 4t \Rightarrow 3t = 12 \Rightarrow t_0 = 4$. Write this value of t in the expressions of x, y, z with t:

$$x_0 = 1 - t_0 = 1 - 4 = -3, y_0 = -16 + 4t_0 = -16 + 16 = 0, z_0 = t_0 = 4.$$

So, the intersection point of the given straight lines is $A(-3;0;4)$.

Now, write the equation of the plane that passes through these intersecting straight lines. As this plane passes through the point $A(-3;0;4)$ and is parallel to the vectors $\vec{a}(-1;4;1)$ and $\vec{b}(-4;-2;1)$, by (1) we can write its equation as follows:

$$\begin{vmatrix} x+3 & y & z-4 \\ -1 & 4 & 1 \\ -4 & -2 & 1 \end{vmatrix} = 0 \Rightarrow$$

$$\Rightarrow 4(x+3)-4y+2(z-4)+16(z-4)+y+2(x+3)=0 \Rightarrow$$

$$\Rightarrow 6x-3y+18z-54=0 \Rightarrow 2x-y+6z-18=0.$$

The last equality is the general equation of the plane that passes through the given straight lines.

Answer: 1) crossing; 3) parallel and are on the plane $2x-y+6z-18=0$; 4) coincide.

Problem 132. Find the angle between the straight lines:

\+ 1) $\frac{x-1}{2}=\frac{y-2}{3}=\frac{z+3}{-1}$ and $\frac{x+1}{-3}=\frac{y}{4}=\frac{z-10}{6}$;

2) $x=5-2t, y=6+4t, z=8t$ and $x=1+t, y=-2t, z=3-4t$.

Answer: 1) $90°$; 2) $180°$. ***Guideline:*** Use formula (1.44).

Problem 133. Find the angle between the plane $4x+4y-7z+1=0$ and the following straight line:

1) $\frac{x-2}{4}=\frac{y-1}{4}=\frac{z+3}{-7}$; 2) $\frac{x-1}{11}=\frac{y+1}{-4}=\frac{z+3}{4}$.

Answer: 1) $90°$; 2) $0°$. ***Guideline:*** Use formula (1.48).

Problem 134*. Derive the formula for the distance from the point $M_1(x_1;y_1;z_1)$ to the given straight line given by the equation $\frac{x-x_0}{a_1}=\frac{y-y_0}{a_2}=\frac{z-z_0}{a_3}$.

Solution. It is clear that the straight line l given by the equation $\frac{x-x_0}{a_1}=\frac{y-y_0}{a_2}=\frac{z-z_0}{a_3}$ passes through the point $M_0(x_0;y_0;z_0)$ and has the direction vector $\vec{a}(a_1,a_2,a_3)$.

Draw a perpendicular from the point M_1 to the straight line l and denote its length by d. It is clear from the figure that for the area of the parallelogram constructed on $\overrightarrow{M_0M_1}$ and $\vec{a}$ we can write $S_{par.} = \left|\vec{a}\right| d$. On the other hand, by definition of the vectorial product, $S_{par.} = \left|\overrightarrow{M_0M_1} \times \vec{a}\right|$. Comparing these expressions obtained for the area of the parallelogram, we get

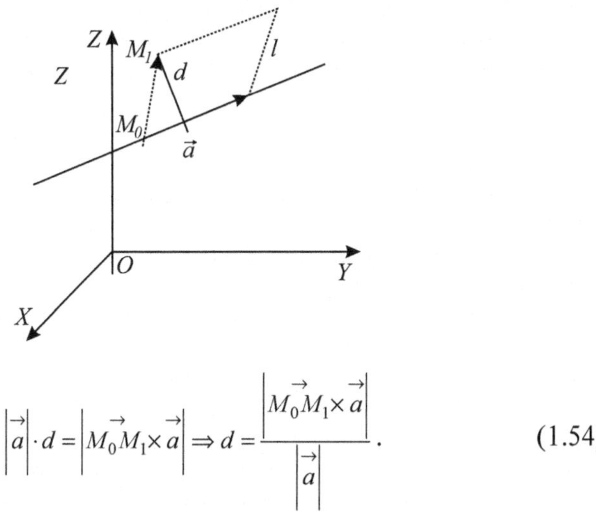

$$\left|\vec{a}\right| \cdot d = \left|\overrightarrow{M_0M_1} \times \vec{a}\right| \Rightarrow d = \frac{\left|\overrightarrow{M_0M_1} \times \vec{a}\right|}{\left|\vec{a}\right|}. \tag{1.54}$$

Equation (1.54) is the formula for the distance from the point on the space to the straight line. If we express this formula by coordinates, we can write

$$d = \frac{\operatorname{mod}\begin{vmatrix} \vec{i} & \vec{j} & \vec{k} \\ x_1 - x_0 & y_1 - y_0 & z_1 - z_0 \\ a_1 & a_2 & a_3 \end{vmatrix}}{\sqrt{a_1^2 + a_2^2 + a_3^2}} =$$

$$= \frac{\sqrt{\begin{vmatrix} y_1 - y_0 & z_1 - z_0 \\ a_2 & a_3 \end{vmatrix}^2 + \begin{vmatrix} z_1 - z_0 & x_1 - x_0 \\ a_3 & a_1 \end{vmatrix}^2 + \begin{vmatrix} x_1 - x_0 & y_1 - y_0 \\ a_1 & a_2 \end{vmatrix}^2}}{\sqrt{a_1^2 + a_2^2 + a_3^2}}. \tag{1.55}$$

Problem 135. Find the distance between the parallel straight lines $\frac{x-2}{1} = \frac{y+1}{2} = \frac{z+3}{2}$ and $\frac{x-1}{1} = \frac{y-1}{2} = \frac{z+1}{2}$.

Answer: $\frac{4\sqrt{2}}{3}$.

Guideline: Find the distance from arbitrary point on one of straight lines to another straight line by formula (1.55).

Problem 136. Write the equation of a straight line that passes through the point $M_0(1;-2;3)$ and forms the angle of 45° with the axis *OX* and the angle of 60° with the axis *OY*.

Answer: $\frac{x-1}{\sqrt{2}}=\frac{y+2}{1}=\frac{z-3}{\pm 1}$. ***Guideline:*** Use formula (1.43).

Home tasks

Problem 137. Write the equation of a plane that passes through the point M_0 and is parallel to noncollinear vectors $\vec{a}$ and $\vec{b}$:

$$1)\ M_0(1;1;1), \vec{a}(0,1,2), \vec{b}(-1,0,1);$$

$$2)\ M_0(0;1;2), \vec{a}(2,0,1), \vec{b}(1,1,0).$$

Answer: $1) x-2y-z=0; 2) -x+y+2z-5=0.$

Problem 138. Write the equation of a plane that passes through the point $M_0(-4;0;4)$ and separates the segments $a=4, b=3$ on the axes *OX* and *OY*.

Answer: $3x+4y+6z-12=0.$

Problem 139. Write the equation of a plane that passes through the point $M_0(2;-3;4)$ and is perpendicular to the vector $\vec{a}=3\vec{i}+2\vec{j}+5\vec{k}$.

Answer: $3x+2y+5z-20=0.$

Problem 140. Write the equation of a plane that passes through the points $M_1(3;2;-1)$, $M_2(1;-1;2)$ and is parallel to the vector $\vec{a}(-2;1;1)$.

Answer: $3x+2y+4z-9=0.$

Problem 141. Write the equation of a plane that passes through the point $A(1;-1;2)$ and is parallel to the plane given by the parametric equations

$x=4-u+v,\ y=2+u+2v,\ z=-1+7u+3v.$

Answer: $11x-10y+3z-27=0$.

Guideline: Write the general equation of the given plane; see solution 2) of 118.

Problem 142. Write the equation of a plane that passes through the point $A(1;1;-1)$ and is perpendicular to the planes $2x-y+5z+3=0$ and $x+3y-z-7=0$.

Answer: $2x-y-z-2=0$.

Problem 143. Find the distance from the point $M_0(5;1;-1)$ to the plane given by the equation $x-2y-2z+4=0$.

Answer: 3.

Problem 144. Write the equation of a plane that passes through the points $M_1(1;3;0)$, $M_2(4;-1;2)$, $M_3(3;0;1)$. Find the distance from the point $M_0(4;3;0)$ to this plane.

Answer: $2x+y-z-5=0$, $\sqrt{6}$.

Problem 145. Find the angle between the planes given by the following equations:

1) $x+2y-z=1$ and $x-y=3$;
2) $x+2y-z-1=0$ and $3x-5y-7z=0$;
3) $x-3y+2z+1=0$ and $6z-9y+3x+5=0$.

Answer: 1) $arc\ \cos\frac{1}{2\sqrt{3}}$; 2) 90°; 3) 0°.

Problem 146. Find the distance between parallel planes with the equations $2x+2y-z+3=0$ and $2x+2y-z+18=0$.

Answer: 5.

Problem 147. Determine mutual state of planes given by general equations:

1) $3x+y-z+1=0$ and $5x+3y+z+2=0$;
2) $x+y-2z+1=0$ and $6z-3x-3y-3=0$;
3) $-x+y+z=1$ and $x-y-z=2$.

Answer: 1) intersect; 2) coincide; 3) parallel.

Guideline: Study the system of linear equations composed of the equations $A_1x+B_1y+C_1z+D_1=0$ and $A_2x+B_2y+C_2z+D_2=0$. For that calculate the ranks of the matrices

$$N_1=\begin{pmatrix} A_1 & B_1 & C_1 \\ A_2 & B_2 & C_2 \end{pmatrix},\ N_2=\begin{pmatrix} A_1 & B_1 & C_1 & D_1 \\ A_2 & B_2 & C_2 & D_2 \end{pmatrix}.$$

For $(rank\, N_1=r_1,\ rank\, N_2=r_2)$ this system has infinitely many solutions, so, the given planes intersect. For $r_1=r_2=2$, the system has no solution, that is, the planes are parallel. For $r_1=r_2=1$ the equations of the system are equipotential, so, the planes coincide.

Problem 148. Write the equation of a plane that intersects the coordinate axes *OX*, *OY*, *OZ* at the points with abscissa *a*, ordinate *b*, applicate *c*. Find the distance from the point $M_0(a;b;c)$ to this plane.

Answer: $\dfrac{2abc}{\sqrt{a^2b^2+a^2c^2+b^2c^2}}$.

Problem 149. Write parametric and canonical equations of a straight line that is the intersection of two planes given by general equations:

$$1)\begin{cases} 2x-y+2z-3=0, \\ x+2y-z-1=0; \end{cases}\quad 2)\begin{cases} x+2y-3z-5=0, \\ 2x-y+z+2=0. \end{cases}$$

Answer: $1)\ x=\frac{7}{5}-\frac{3}{5}t,\ y=-\frac{1}{5}+\frac{4}{5}t,\ z=t,\ \dfrac{x-\frac{7}{5}}{-3}=\dfrac{y+\frac{1}{5}}{4}=\dfrac{z}{5};$

$2)\ x=\frac{1}{5}+\frac{1}{5}t,\ y=\frac{12}{5}+\frac{7}{5}t,\ z=t,\ \dfrac{x-\frac{1}{5}}{1}=\dfrac{y-\frac{12}{5}}{7}=\dfrac{z}{5}.$

Problem 150. Write the equation of a straight line that passes through the given two points M_1 and M_2:

$1)\ M_1(1;-2;1),\ M_2(3;1;-1);\ 2)\ M_1(3;-1;0),\ M_2(1;0;-3);$

$3)\ M_1(3;-2;1),\ M_2(3;1;-1);\ 4)\ M_1(-1;2;3),\ M_2(5;2;3).$

Answer: $1)\ \dfrac{x-1}{2}=\dfrac{y+2}{3}=\dfrac{z-1}{-2};\ 2)\ \dfrac{x-3}{-2}=\dfrac{y+1}{1}=\dfrac{z}{-3};$

$$3)\, x=3, \frac{y+2}{3}=\frac{z-1}{-2};\ 4)\, y=2, z=3.$$

Problem 151. Verify if the given three points are on the straight line. If they are not on one straight line, write the equation of the plane that passes through these three points:

$$1)\, A(3;0;0), B(0;-1;0), C(0;0;4);$$
$$2)\, A(2;1;1), B(2;0;-1), C(2;4;3);$$
$$3)\, A(1;1;1), B(0;-1;2), C(2;3;-1);$$
$$4)\, A(1;-1;2), B(3;2;1), C(-1;-4;3).$$

Answer: 1) $4x-12y+3z-12=0$; 2) $x=2$;

3) $2x-y-1=0$; 4) are on one straight line.

Problem 152. Write the equation of a straight line of the plane $x-3y+2z+1=0$ that passes through the intersection points of the straight lines

$$\frac{x-5}{5}=\frac{y+1}{-2}=\frac{z-3}{-1} \text{ and } \frac{x-3}{4}=\frac{y+4}{-6}=\frac{z-5}{2}.$$

Answer: $\frac{x+1}{7}=\frac{y-2}{-1}=\frac{z-3}{-5}$.

Problem 153. The point A is on the straight line $\frac{x-1}{2}=\frac{y}{3}=\frac{z+1}{1}$ and the distance from this point to the plane $x+y+z+3=0$ is $\sqrt{3}$. Find the coordinates of the point A.

Answer: (1;0; – 1) or (– 1; – 3; – 2).

Problem 154. Write the equation of a straight line that passes through the point $M(1;1;1)$ and is perpendicular to the vectors $\vec{a}_1 = 2\vec{i}+3\vec{j}+\vec{k},\ \vec{a}_2 = 3\vec{i}+\vec{j}+2\vec{k}$.

Answer: $\frac{x-1}{5}=\frac{y-1}{-1}=\frac{z-1}{-7}$.

Guideline: Take into account that the vector being the vectorial product of $\vec{a}_1$ and $\vec{a}_2$ is a direction vector of this straight line.

Problem 155. Find the distance from the point $M(2; -1;3)$ to the straight line $\frac{x+1}{3}=\frac{y+2}{4}=\frac{z-1}{5}$.

Answer: $0,3\sqrt{38}$.

Guideline: Use formula (1.55) in problem 134.

Problem 156. Determine how the straight line given by the following equation was arranged with respect to the plane $x-3y+z+1=0$:

$$1)\ \frac{x+1}{1}=\frac{y+1}{5}=\frac{z}{2};\quad 2)\begin{cases}3x-2y-1=0,\\ 7y-3z-4=0.\end{cases}$$

Answer: 1) the straight line intersects the plane at the point $\left(-\frac{3}{4},\frac{1}{4},\frac{1}{2}\right)$; 2) the straight line is on the plane.

Problem 157. Determine mutual state of straight lines given by the following equations in space:

$$1)\ \frac{x-1}{2}=\frac{y+2}{-3}=\frac{z-5}{4}\ \text{and}\ \frac{x-7}{3}=\frac{y-2}{2}=\frac{z-1}{-2};$$

$$2)\ \frac{x+3}{2}=\frac{y-2}{7}=\frac{z+4}{-11}\ \text{and}\ \frac{x-1}{2}=\frac{y+3}{2}=\frac{z-6}{-1};$$

$$3)\ \frac{x-8}{1}=\frac{y-5}{2}=\frac{z-8}{3}\ \text{and}\ \frac{x-9}{2}=\frac{y-7}{4}=\frac{z-3}{6}.$$

Answer: 1) intersect; 2) crossing; 3) parallel.

Problem 158. Find the angle between the plane $4x+4y-7z+1=0$ and the straight line $\frac{x-1}{3}=\frac{y+2}{2}=\frac{z}{-6}$.

Answer: $arc\sin\frac{62}{63}$.

Problem 159. Prove that the straight lines $\frac{x}{2}=\frac{y}{3}=\frac{z}{1}$ and $x=z+1, y=1-z$ are mutually perpendicular.

Problem 160. Find the angle between the straight lines $\begin{cases}2x+y-z+1=0,\\ x+3y+z+2=0\end{cases}$ and $\begin{cases}x+3y-z+2=0,\\ x+y+z-1=0\end{cases}$.

Answer: $\arccos\frac{\sqrt{3}}{5}$.

Problem 161. Write the canonical equation of a straight line that passes through the point $M_0(3;-2;-4)$ and is parallel to the plane $3x - 2y - 3z - -7 = 0$ and intersect the straight line $\frac{x-2}{3} = \frac{y+4}{-2} = \frac{z-1}{2}$.

Answer: $\frac{x-3}{5} = \frac{y+2}{-6} = \frac{z+4}{9}$.

Guideline: It is necessary to write the equation of the straight line in the parametric form, to write the expressions obtained for x,y,z $(x = 3t+2, y = -2t-4, z = 2t+1)$ in the canonical equation $\frac{x-3}{a_1} = \frac{y+2}{a_2} = \frac{z+4}{a_3} = t$ of the sought-for straight line and to take into account the expressions of a_1, a_2, a_3 found with t in the equality $3a_1 - 2a_2 - 3a_3 = 0$ that is the condition of parallelism of the straight line to plane $3x - 2y - 3z - 7 = 0$.

1.12 SECOND-ORDER CURVES

When at least one of a_{11}, a_{12}, a_{22} is non-zero, *a, b, c* are any known numbers, the points set whose coordinates satisfy the equation

$$a_{11}x^2 + 2a_{12}xy + a_{22}y^2 + ax + by + c = 0, \tag{1.56}$$

on a surface, is called a second-order curve.

In this section, we will be acquainted with the simplest kinds of second-order curves as a circle, ellipse, hyperbola, and parabola.

Definition. Circle is the focus of points equidistant from a point called, the center of the plane.

If the center of a circle is on the point *C*(*a*, *b*), we can denote any point on this circle by *M*(*x*, *y*) and by definition we can write

$$|CM| = r = const. \tag{1.57}$$

r is the radius of a circle. If we denote eq. (1.57) by coordinates, we get

$$\sqrt{(x-a)^2 + (y-b)^2} = r \Rightarrow (x-a)^2 + (y-b)^2 = r^2. \tag{1.58}$$

Equation (1.58) is the canonic equation of r radius circle with the center at the point $C(a, b)$.

In the special case, if the center of a circle is at the origin of coordinates, in eq. (1.58) we write $a = b = 0$ and get

$$x^2 + y^2 = r^2 \tag{1.59}$$

Equation (1.59) is the equation of r radius circle with the center at the origin of coordinates. In the parametric form, this circle is given by the equations $x = r\cos t, y = r\sin t \ (t \in [0, 2\pi))$.

Having opened the parenthesis in the left-hand side of eq. (1.58), we write it in the form eq. (1.56):

$$x^2 + y^2 - 2ax - 2by + \left(a^2 + b^2 - r^2\right) = 0. \tag{1.60}$$

It is seen from eq. (1.60) that for the equation in the form eq. (1.56) to be a circle equation, in eq. (1.56) it is necessary $a_{12} = 0$ and $a_{11} = a_{22}$. But not all of the equalities in the form

$$x^2 + y^2 + Ax + By + C = 0 \tag{1.61}$$

are circle equations. To verify if the equality in the form eq. (1.61) is a circle equation, we must expand perfect squares of binomials in its left-hand side with respect to x and y:

$$(1.61) \Rightarrow \left(x - \frac{A}{2}\right)^2 + \left(y - \frac{B}{2}\right)^2 = \frac{A^2 + B^2 - 4C}{4} \tag{1.62}$$

It is clear that for $A^2 + B^2 - 4C > 0$ eqs. (1.62) and (1.61) are circle equations.

For $A^2 + B^2 - 4C = 0$, only the coordinates of the point $\left(\frac{A}{2}; \frac{B}{2}\right)$ satisfy eq. (1.62) on the plane.

For $A^2 + B^2 - 4C < 0$, there is no point whose coordinates satisfy eq. (1.62) on the plane.

The equation of the straight line tangent to this circle at the point $M_0(x_0, y_0)$ on r radius circle is in the form

$$xx_0 + yy_0 = r^2. \tag{1.63}$$

The straight line tangent to this circle at the point $M_0(x_0, y_0)$ on the circle given by eq. (1.58) is given by the equation

$$(x-a)(x_0-a)+(y-b)(y_0-b)=r^2. \tag{1.64}$$

The sufficient and necessary condition for the straight line given by the equation $Ax+By+C=0$ be a straight line tangent to the circle possessing eq. (1.58) is satisfaction of the relation

$$|Aa+Bb+C|=r\sqrt{A^2+B^2} \tag{1.65}$$

Definition. Ellipse is a set of all points in a plane for which the sum of distances from two points called a focus is constant and greater than the distance between the focal points.

If $F_1(-C; 0)$ and $F_2(-C; 0)$, $(C > 0)$ are the focal points, $M(x, y)$ is any point on an ellipse, by definition of an ellipse we can write

$$|MF_1|+|MF_2|=2a=const,\ (2a>2c). \tag{1.66}$$

If we express eq. (1.66) by coordinates, after certain transformations we get the equation

$$\frac{x^2}{a^2}+\frac{y^2}{b^2}=1.$$

Here, b denotes a positive number satisfying the relation $a^2 - c^2 = b^2$.

Equation (1.66) is said to be a canonical equation of the ellipse.

From canonic eq. (1.66) of the ellipse we get its following properties:

1. An ellipse intersects the abscissa axis at the two points $A_1(-a; 0)$ and $A_2(-a; 0)$:

$$y=0,\ \frac{x^2}{a^2}=1\Rightarrow x=\pm a;$$

2. The ellipse intersects the ordinate axis at the two points $B_1(0; -b)$ and $B_2(0; -b)$:

$$x=0,\ \frac{y^2}{b^2}=1\Rightarrow y=\pm b;$$

3. If the point $M_1(x, y)$ is on the ellipse, then the points $M_2(-x, y)$, $M_3(x, -y)$ and $M_4(-x, -y)$ are on ellipse as well. This means that an ellipse is symmetric both with respect to coordinate axis and origin of coordinates.

$2a$ is said to be the length of the great axis of an ellipse, $2b$ the length of the small axis. a is called a semi-major axis, b a semi-minor axis.

For $a = b = r$, an ellipse is a circle given by eq. (1.59).

The distances $r_1 = |F_1M|$, $r_2 = |F_2M|$ from the point $M(x, y)$ on an ellipse to the focal points F_1 and F_2 are called focal radii of the point M.

Focal radii may be calculated by the formulas

$$r_1 = a + \frac{c}{a}x, \ r_2 = a - \frac{c}{a}x. \tag{1.67}$$

The number $e = \frac{c}{a}$ is called eccentricity of the ellipse.

As $c < a$, the eccentricity of the ellipse is less than a unit.

The ellipse with semi-axis a and b may be given by the parametric equation $x = a\cos t, y = b\sin t, \frac{t \in (0.2\pi)}{t \in (0.2\pi)} t \in (0.2\pi)$.

The equation of the straight line tangent to this ellipse at the point $M_0(x_0, y_0)$ is written as follows

$$\frac{xx_0}{a^2} + \frac{yy_0}{b^2} = 1. \tag{1.68}$$

The necessary and sufficient condition for the straight line with the general equation

$$Ax + By + C = 0 \tag{1.69}$$

to be a tangential straight line to the ellipse given by eq. (1.66) is satisfaction of the relation

$$A^2 a^2 + B^2 b^2 = C^2 \tag{1.70}$$

Definition. A set of points in a plane for which the difference of modulus of distances from two points called a focus is constant and less than the distance between the focal points is called a hyperbola.

If we denote the focal points by $F_1(-c, 0)$, $F_2(c, 0)$, $(c > 0)$, arbitrary point on a hyperbola by $M(x, y)$, by definition we can write

$$\left||MF_1|-|MF_2|\right| = 2a,\ (2a < 2c)\,. \tag{1.71}$$

If we express eq. (1.71) by coordinates, after certain transformations we get the equation

$$\frac{x^2}{a^2} - \frac{y^2}{b^2} = 1\,. \tag{1.72}$$

Here, b denotes a positive number satisfying the condition $c^2 - a^2 = b^2$.

Equation (1.72) is said to be a canonical equation of the hyperbola.

The following properties of a hyperbola are obtained from its canonical equation:

1. Hyperbola intersects the abscissa axis at two points $A_1(-a; 0)$ and $A_2(-a; 0)$:
$$y = 0,\ \frac{x^2}{a^2} = 1 \Rightarrow x = \pm a;$$
2. A hyperbola does not intersect the ordinate axis;
3. A hyperbola is symmetric both with respect to coordinate axis and origin of coordinates.

a is said to be a real semi-axis, b an imaginary semi-axis.

As c > a, the eccentricity $e = \dfrac{c}{a}$ of the hyperbola is greater than a unit.

The straight lines with the equations

$$y = \frac{b}{a}x,\ y = -\frac{b}{a}x \tag{1.73}$$

are called asymptotes of the hyperbola given by eq. (1.71).

The equation of a straight line at the point $M_0(x_0, y_0)$ arranged on the hyperbola given by eq. (1.72), tangent to this hyperbola is written as

$$\frac{xx_0}{a^2} - \frac{yy_0}{b^2} = 1\,. \tag{1.74}$$

The necessary and sufficient condition for the straight line with eq. (1.69) to be tangent to the hyperbola given by eq. (1.72), is satisfaction of the relation

$$A^2a^2 - B^2b^2 = C^2. \qquad (1.75)$$

Definition. A parabola is the set of all points on the plane equidistant from the given point called a focus and the straight line called a directrix.

Denote the distance between the focal point and matrix by F. If we take the origin of coordinates at the midplane of the perpendicular drawn to the directrix at the focal point F, it is clear that $F\left(\frac{p}{2},0\right)$. The directrix is the straight line $x=-\frac{p}{2}$. Taking on a parabola an arbitrary point $M(x, y)$, draw from M to the directrix the perpendiculars MN, $\left(N\left(-\frac{p}{2},y\right)\right)$. By definition of parabola

$$|MF| = |MN| \Rightarrow \sqrt{\left(x-\frac{p}{2}\right)^2 + y^2} =$$

$$= \sqrt{\left(x+\frac{p}{2}\right)^2 + 0^2} \Rightarrow x^2 - px + \frac{p^2}{4} + y^2 =$$

$$= x^2 + px + \frac{p^2}{4} \Rightarrow y^2 = 2px. \qquad (1.76)$$

Equation (1.76) is said to be a canonical equation of the parabola.

From canonical eq. (1.76) we get the following properties of parabola:

1. If the point $M_1(x, y)$ is on the parabola, the point $M_2(x, -y)$ is also on the parabola. This means that this parabola is symmetric with respect to the abscissa axis;
2. The coordinates of the point $O(0;0)$ satisfy eq. (1.76). Thus, the parabola passes through the origin of coordinates;
3. As we get $x=\frac{y^2}{2p}\geq 0$ from eq. (1.76), the abscissas of all points on the parabola are not negative, that is, except the vertex point all the points of the parabola are in the right hand side of the axis OY.

The equation of a parabola with the focal point at the point $F\left(-\frac{p}{2},0\right)$ is written as

$$y^2 = -2px, \qquad (1.77)$$

the equation of the parabola with the focal point $F\left(0;\frac{p}{2}\right)$ as

$$x^2 = 2py, \tag{1.78}$$

the equation of the parabola with the focal point $F\left(0;-\frac{p}{2}\right)$ as

$$x^2 = -2py. \tag{1.79}$$

At the point (x_0, y_0) on the parabola $y^2 = 2px$ this equation of a tangent to the parabola, is given by the formula:

$$yy_0 = p(x + x_0). \tag{1.80}$$

The necessary and sufficient condition for the straight line $Ax + By + C = 0$ to be a straight line tangential to the parabola $y^2 = 2px$ is satisfaction of the relation

$$pB^2 = 2AC. \tag{1.81}$$

When the second-order curve is given by general eq. (1.56), the equation of the straight line tangential to this curve at the point $M_0(x_0, y_0)$ on this plane may be written as follows

$$a_{11}xx_0 + a_{12}(xy_0 + x_0y) +$$

$$+a_{22}yy_0 + a(x + x_0) + b(y + y_0) + c = 0. \tag{1.82}$$

Problems to be solved on auditorium

Problem 162. Find the coordinates and radius of the center of the circle given by the following equation:

$$1)\, x^2 + y^2 + 4y = 0;\ 2)\, 7x^2 + 7y^2 - 2x - 7y - 1 = 0.$$

Answer: $1)(0;-2), 2;\ 2)\left(\frac{1}{7};\frac{1}{2}\right), \frac{9}{14}.$

Problem 163. Write the canonical equation of a circle of radius $r = 7$, centered at the point $C(2; -3)$.

Answer: $(x-2)^2 + (y+3)^2 = 49.$

Problem 164. The point $M(2;6)$ is on a circle centered at the point $C(-1;2)$. Write the canonical equation of this circle.

Answer: $(x+1)^2+(y-2)^2=25.$

Problem 165. Write the equation of the circle of radius 5, centered at the origin of coordinates. Write the equation of the straight line on this plane and that is tangent to this circle at the point with abscissa 3.

Answer: $x^2+y^2=25$, $3x+4y=25$, and $3x-4y=25$.

Problem 166. Write the equation of the straight line tangential to the circle $(x-1)^2+(y+2)^2=25$ at the point M(– 3;1).

Answer: $4x-3y+15=0.$

Guideline: Use eq 1.64.

Problem 167. Show that the circles $(x+1)^2+(y-1)^2=45$ and $(x+1)^2+(y-5)^2=5$ are tangential. Write the equation of the common tangent passing through this tangential point of these circles.

Answer: $x+2y-16=0.$

Problem 168. Find the lengths, coordinates, and eccentricity of local points of the ellipse $9x^2+25y^2=225$.

Answer: $a=5,\ b=3,\ F_1(-4;0),\ F(4;0),\ e=\frac{4}{5}.$

Problem 169. Write the canonical equation of the ellipse by the following data:

a) the semi-major axis $a=5$, the distance between the focal points is 8;

b) the distance between the focal points equals 6, eccentricity is $\frac{3}{5}$.

Answer: *a)* $\frac{x^2}{25}+\frac{y^2}{9}=1;$ *b)* $\frac{x^2}{25}+\frac{y^2}{16}=1.$

Problem 170. We are given the ellipse $25x^2+144y^2=1$. Determine if the points $A\left(1;\frac{1}{6}\right)$, $B\left(\frac{1}{13},\frac{1}{13}\right)$, $C\left(\frac{1}{6},-\frac{1}{24}\right)$ are on the inside or outside of the ellipse.

Answer: *A* is in outside, *B* is on it, *C* is inside.

Problem 171. Write the equation of the straight line tangential to the ellipse $\frac{x^2}{12}+\frac{y^2}{4}=1$ at the point (3;1).

Answer: $x+y-4=0.$

Guideline: Use eq. (1.68).

Problem 172. Determine if the given straight lines given by the following equations are tangent to the ellipse $\frac{x^2}{48}+\frac{y^2}{36}=1$. If they are tangent find the coordinates of tangency point:

1) $x + y - 4 = 0$; 2) $3x - 2y - 24 = 0$.

Answer: 1) is not tangent; 2) is tangent at the point $(6; -3)$.

Guideline: verify if condition (1.70) is satisfied. If this condition is satisfied, solve these equations jointly.

Problem 173. Write the equation of such a tangent to the given ellipse given by the equation $\frac{x^2}{4}+\frac{y^2}{9}=1$ that,

1) it be parallel to the straight line $x - 2y + 1 = 0$;
2) be perpendicular to the straight line $x - 2y + 1 = 0$.

Solution of 1): If we denote the coordinates of the tangency point of the sought-for tangent to the given ellipse by x_0, y_0, then equation of this tangent is in the form

$$\frac{xx_0}{4}+\frac{yy_0}{9}=1 \Rightarrow (9x_0)x+(4y_0)y-36=0.$$

According to the condition, this tangent must be parallel to the straight line given by the equation $x - 2y + 1 = 0$. By the condition of parallelism of two straight lines given by common equations (see eq. (1.69) in 1.12), we can write:

$$\frac{9x_0}{1}=\frac{4y_0}{-2} \Rightarrow 9x_0=-2y_0 \Rightarrow x_0=-\frac{2}{9}y_0.$$

As the tangency point (x_0, y_0) is on the ellipse, the coordinates of this point should satisfy the equation of the ellipse:

$$\frac{x_0^2}{4}+\frac{y_0^2}{9}=1 \Rightarrow 9x_0^2+4y_0^2=36.$$

Take into account the expression of x_0 obtained above by y_0 in the last equality:

$$9\cdot\frac{4}{81}y_0^2+4y_0^2=36\Rightarrow\frac{4}{9}y_0^2+4y_0^2=36\Rightarrow$$

$$\Rightarrow 4y_0^2+36y_0^2=36\cdot 9\Rightarrow 40y_0^2=36\cdot 9\Rightarrow$$

$$\Rightarrow y_0^2=\frac{81}{10}\Rightarrow y_0=\pm\frac{9}{\sqrt{10}}.$$

As $x_0=-\frac{2}{9}y_0$ we get $x_0=-\frac{2}{9}\cdot\left(\pm\frac{9}{\sqrt{10}}\right)=\mp\frac{2}{\sqrt{10}}$.

Thus, we find two values for (x_0, y_0):

$$1)\ x_0=\frac{2}{\sqrt{10}},\ y_0=-\frac{9}{\sqrt{10}};\ 2)\ x_0=-\frac{2}{\sqrt{10}},\ y_0=\frac{9}{\sqrt{10}}.$$

If we substitute these values in the equation $(9x_0)x+(4y_0)y-36=0$ of the tangent, we get

$$\frac{18}{\sqrt{10}}x-\frac{36}{\sqrt{10}}y-36=0\Rightarrow x-2y-2\sqrt{10}=0,$$

$$-\frac{18}{\sqrt{10}}x+\frac{36}{\sqrt{10}}y-36=0\Rightarrow x-2y+2\sqrt{10}=0.$$

So, the ellipse $\frac{x^2}{4}+\frac{y^2}{9}=1$ has two tangents parallel to the straight line $x-2y+1=0$:

$$x-2y-2\sqrt{10}=0 \text{ and } x-2y+2\sqrt{10}=0.$$

Answer: 2) $2x+y-5=0$ and $2x+y+5=0$.

Problem 174. The straight lines $y=\pm 2x$ are the asymptotes of the hyperbola. The point $\left(-\frac{5}{4};\frac{3}{2}\right)$ is on this hyperbola. Write the canonical equation of this hyperbola.

Answer: $x^2-\frac{y^2}{4}=1$.

Problem 175. Write the canonical equation of the hyperbola by the following data:

1) the distance between the focal points is $2c = 10$, the distance between the vertices is $2a = 8$;
2) real semi-axis equals $a = 2\sqrt{5}$, the eccentricity equals $e = \sqrt{1{,}2}$.

Answer: 1) $\frac{x^2}{16} - \frac{y^2}{9} = 1$; 2) $\frac{x^2}{20} - \frac{y^2}{4} = 1$.

Problem 176. Find a point on the hyperbola $9x^2 - 16y^2 = 144$ such that the distance from this point to the left focal point be twice less than the distance to the right focal point.

Solution. At first, we write the equation of the hyperbola in the canonical form: $\frac{x^2}{16} - \frac{y^2}{9} = 1$. It is clear from this equation that $a^2 = 16, b^2 = 9$. Therefore $c^2 = a^2 + b^2 = 16 + 9 = 25 \Rightarrow c = 5$.

The distance from the point $M(x, y)$ on the hyperbola to the left focal point is determined by the formula $|F_1M| = -\frac{c}{a}x - a$, to the right focal point by the formula $|F_2M| = -\frac{c}{a}x + a$. Hence and from the condition of the problem we get:

$$2|F_1M| = |F_2M| \Rightarrow 2 \cdot \left(-\frac{5}{4}x - 4 \right) = -\frac{5}{4}x + 4 \Rightarrow$$

$$\Rightarrow -10x - 32 = -5x + 16 \Rightarrow x = -\frac{48}{5} = -9{,}6.$$

Taking into account this value found for x in the equation of the hyperbola and find

$$9 \cdot \frac{48^2}{25} - 16y^2 = 144 \Rightarrow y^2 = \frac{9 \cdot 48^2}{16 \cdot 25} - \frac{144}{16} = \frac{9 \cdot 3 \cdot 48}{25} - 9 =$$

$$= \frac{9 \cdot 3 \cdot 48 - 25 \cdot 9}{25} = \frac{9 \cdot (144 - 25)}{25} = \frac{9 \cdot 119}{25} \Rightarrow y = \pm \frac{3\sqrt{119}}{5}.$$

Thus, on the plane there are two points satisfying the problem condition:

$$\left(-9{,}6; \frac{3\sqrt{119}}{5} \right) \text{ and } \left(-9{,}6; -\frac{3\sqrt{119}}{5} \right).$$

Problem 177. Determine if the straight line with the equation $\frac{x^2}{20}-\frac{y^2}{36}=1$ is tangent to the hyperbola given by the equation $3x - y - 12 = 0$. If it is tangential, find the coordinates of the tangency point.

Answer: It is tangent at the point (5;3).

Guideline: Verify if eq. (1.75) is satisfied.

Problem 178. Write the equation of the tangent of the hyperbola $4x^2 - 9y^2 = 36$ perpendicular to the straight line $x + 2y = 0$.

Answer: $y = 2x + 4\sqrt{2}$, $y = 2x - 4\sqrt{2}$.

Problem 179. Write the equation of the tangent of the hyperbola $\frac{x^2}{16}-\frac{y^2}{64}=1$ parallel to the straight line $10x - 3y + 9 = 0$.

Answer: $10x - 3y - 32 = 0$, $10x - 3y + 32 = 0$.

Problem 180. Write the canonical equation of a hyperbola by the following data:

1) Passes through the points (0;0), (1;–3) and is symmetric with respect to the abscissa axis;
2) passes through the points (0;0), (2;–4) and is symmetric with respect to the ordinate axis;
3) its vertex is at the origin of coordinates, focal point is at the point $F(0;-3)$.

Answer: 1) $y^2 = 9x$; 2) $x^2 = -y$; 3) $x^2 = -12y$.

Problem 181. Find on the parabola $y^2 = 8x$ a point such that the distance from this point to the directrix of this parabola be equal to 4.

Solution. The directrix of the parabola $y^2 = 2px$ is the straight line $x=-\frac{p}{2}$. In this example, as $y^2 = 8x$, then $p = 4$, the directrix is the straight line $x = -2$. Let the point sought on the parabola be (x, y). Then since the distance from this point to the directrix $x = -2$ is expressed by the formula

$$\sqrt{(x+2)^2+(y-y)^2}=|x+2|$$

as the distance between the two points (x, y) and $(-2; y)$, by the problem condition we can write $|x+2|=4$. Hence

$$|x+2|=4 \Rightarrow x+2=\pm 4 \Rightarrow \left[\begin{array}{l} x+2=4 \\ x+2=-4 \end{array}\right., \Rightarrow \left[\begin{array}{l} x=2, \\ x=-6. \end{array}\right.$$

On the parabola, there is no a point with negative abscissa. Thus, we should take only $x = 2$. We can find the coordinates of the points with the abscissa on the parabola, from the equation of the parabola:

$$y^2 = 8 \cdot 2 = 16 \Rightarrow y = -4,\ y = 4.$$

Thus, distance from both points (2;4) and (2; – 4) on the parabola to the directrix will equal 4.

Problem 182. Write the equation of a straight line tangent to it at the point $M_0\ (x_0; y_0)$ on the parabola:

$$1)\ y^2 = 32x,\ M_0\left(\frac{1}{2};4\right);\ 2)\ x^2 = -12y,\ M_0\left(2;-\frac{1}{3}\right).$$

Answer: 1) $4x - y + 2 = 0$; 2) $x + 3y - 1 = 0$.

Guideline: The equation of the straight line tangent to it at the point $M_0\ (x_0; y_0)$ on the parabola $x^2 = -2\,py$ is in the form $xx_0 = -p(y + y_0)$.

Problem 183. Write the equation of the tangent drawn to the parabola $y^2 = 16x$ from the point (1;5).

Answer: $x - y + 4 = 0$, $4x - y + 1 = 0$.

Problem 184. Determine if the straight line $x + y + 1 = 0$ is tangential to the parabola $y^2 = 4x$. If it is tangential, find the coordinates of the tangency point.

Answer: Is tangential at the point (1;–2).

Problem 185. Write the equation of the tangent parallel to the straight line $x + y = 0$ of the parabola given by the equation $y^2 = 8x$.

Answer: $x + y + 2 = 0$.

Problem 186. Write the equation of the tangent perpendicular to the straight line $2x + y - 4 = 0$ of the parabola given by the equation $y^2 = 10x$.

Answer: $x - 2y + 10 = 0$.

Problem 187. Determine the kind of the second-order curve whose equation is given in polar coordinates and write its canonical equation:

$$1)\ r = \frac{9}{5 - 4\cos\varphi};\ 2)\ r = \frac{9}{4 - 5\cos\varphi};\ 3)\ r = \frac{3}{1 - \cos\varphi}.$$

Solution of 1). If we take the right focal points of ellipse and hyperbola and the focal point of a parabola as a pole, and the focal axis as a polar axis, each of these three curves may be determined in polar coordinates by the following equation

$$r = \frac{p}{1 - e\cos\varphi}. \tag{1.83}$$

Here e denotes the eccentricity of the curve, for $e < 1$, eq. (1.83) is the equation of an ellipse, for $e > 1$, is the equation of a hyperbola.

When eq. (1.83) is the equation of an ellipse or hyperbola, we take $p = \frac{b^2}{a}$.

For determining what curve gives the equation $r = \frac{9}{5 - 4\cos\varphi}$ given in this example at first we reduce it to the form of eq. (1.83).

$$r = \frac{9}{5 - 4\cos\varphi} \Rightarrow r = \frac{\frac{9}{5}}{1 - \frac{4}{5}\cos\varphi}.$$

Comparing the last equation with eq. (1.83), we get $e = \frac{4}{5}$ as $e < 1$, this is the equation of an ellipse. As for the ellipse

$$p = \frac{b^2}{a},\ e = \frac{c}{a},\ a^2 - c^2 = b^2$$

we jointly solve the equations

$$\frac{b^2}{a} = \frac{9}{5},\ \frac{c}{a} = \frac{4}{5},\ a^2 - c^2 = b^2,$$

find a and b :

$$b^2 = \frac{9}{5}a,\ c^2 = \frac{16}{25}a,\ a^2 = \frac{16}{25}a^2 = \frac{9}{5}a \Rightarrow$$

$$\Rightarrow a = 5,\ b^2 = \frac{9}{5}\cdot 5 = 9\ .$$

Thus, the given equality is the equation of an ellipse and the canonical equation of this ellipse is in the form

$$\frac{x^2}{25} + \frac{y^2}{9} = 1\ .$$

Answer: 2) $\frac{x^2}{16} - \frac{y^2}{9} = 1$; 3) $y^2 = 6x$.

Problem 188*. Write the following equations in polar coordinates. Take as a pole the right focal point, as a polar axis take a focal axis:

1) $\frac{x^2}{16} - \frac{y^2}{9} = 1$; 2) $\frac{x^2}{25} - \frac{y^2}{16} = 1$; 3) $y^2 = 6x$.

Solution of 1): For writing eq. (1.83) from the polar coordinates, we should find the parameters e and p. Taking into account from the equation of the given hyperbola $a = 4$, $b = 3$ and $b^2 = c^2 - a^2$ for the hyperbola, we get:

$$c^2 = a^2 + b^2 = 16 + 9 = 25 \Rightarrow c = 5\cdot$$

It is known that $e = \frac{c}{a}$ and $p = \frac{b^2}{a}$ (see the solution of 187, 1))
In eq. (1.83) take into account the values $e = \frac{5}{4}$ and $p = \frac{9}{4}$:

$$r = \frac{p}{1 - e\cos\varphi} \Rightarrow r = \frac{\frac{9}{4}}{1 - \frac{5}{4}\cos\varphi} \Rightarrow r = \frac{9}{4 - 5\cos\varphi}\ .$$

The last equality is the equation of the given hyperbola in polar coordinates.

Answer: 2) $r = \frac{16}{5 - 3\cos\varphi}$; 3) $r = \frac{3}{1 - \cos\varphi}$.

Problem 189. Write the canonic equations of the curves given by the following parametric equations:

1) $x = t^2 - 2t + 1,\ y = t - 1,\ t \in (-\infty, +\infty)$;
2) $x = 3\cos t,\quad y = 2\sin t,\quad t \in (0.2\pi)$;
3) $x = -1 + 2\cos t,\quad y = 3 + 2\sin t,\quad t \in (0.2\pi)$;
4) $x = 4\sec t,\ y = tgt,\ t \in \left(-\dfrac{\pi}{2}; \dfrac{\pi}{2}\right)$.

Answer: 1) $y^2 = x$; 2) $\dfrac{x^2}{9} + \dfrac{y^2}{4} = 1$;

3) $(x+1)^2 + (y-3)^2 = 4$; 4) $\dfrac{x^2}{4} - y^2 = 1$.

Guideline: Annihilate the parameter t from the equations.

Home tasks

Problem 190. Find the coordinates and radius of the center of the circle given by the following equation:

1) $x^2 + y^2 + 5x - 5y + 12 = 0$
2) $2x^2 + 2y^2 - 12x + y + 3 = 0$.

Answer: 1) $(-2,5;2,5), \dfrac{1}{\sqrt{2}}$; 2) $\left(3; -\dfrac{1}{4}\right), \dfrac{11}{4}$.

Problem 191. Write the canonical equation of the circle of radius $r = 5$, centered at the point $C(-1;2)$.

Answer: $(x+1)^2 + (y-2)^2 = 25$.

Problem 192. The point $M(8;7)$ is on the circle centered at the point $C(2; -1)$. Write the canonical equation of this circle.

Answer: $(x-2)^2 + (y+1)^2 = 100$.

Problem 193. Write the equation of the circle of radius 10 and centered at the origin of coordinates. Write the equation of the straight line that is on this circle and tangent to this circle at the point with ordinate 6.

Answer: $x^2 + y^2 = 100$, $4x + 3y - 50 = 0$, $4x - 3y + 50 = 0$.

Problem 194*. Write the equation of the straight line that passes through the point $M(1;4)$ and is tangent to the circle $(x-1)^2 + (y+1)^2 = 9$.

Answer: $4x + 3y - 16 = 0$, $4x - 3y + 8 = 0$.

Guideline: Denoting the tangency point of the straight line by (x_0, y_0), use the following relations:

$$(x_0 - 1)^2 + (y_0 + 1)^2 = 9, \ (x_0 - 1)(x - 1) + (y_0 + 1)(y + 1) = 9,$$

$$\frac{x-1}{x_0 - 1} = \frac{y-4}{y_0 - 4}.$$

Problem 195. Show that $(x-1)^2 + (y-2)^2 = 18$ the circles and $(x-5)^2 + (y-6)^2 = 2$ are tangent to each other. Write the equation of the common tangent passing through this tangency point of these circles.

Answer: $x + y - 9 = 0$.

Problem 196. Find the length, focal points, and eccentricities of the semi-axis of the ellipse $x^2 + 4y^2 = 1$.

Answer: $a = 1, b = \frac{1}{2}, F_1\left(0;\frac{\sqrt{3}}{2}\right), F_2\left(0;-\frac{\sqrt{3}}{2}\right), e = \frac{\sqrt{3}}{2}$.

Problem 197. The ellipse $9x^2 + 16y^2 = 2$ is given. Determine if the points $A\left(\frac{1}{3};-\frac{1}{4}\right)$, $B(1;2)$, $C\left(\frac{1}{7};\frac{2}{7}\right)$ are on, inside or outside of the ellipse.

Answer: A is on, B is outside, and C is inside.

Problem 198. Write the equation of the straight line tangent to it at the point $M_0\left(\frac{1}{3};-\frac{1}{4}\right)$ on the ellipse $9x^2 + 16y^2 = 2$.

Answer: $3x - 4y - 2 = 0$.

Problem 199. Write the equation of the tangent parallel to the secant of the first coordinate quarter of the ellipse $x^2 + 4y^2 = 20$.

Answer: $x - y = \pm 5$.

Problem 200. Write the equation of the tangent that passes through the point (0;6) of the ellipse $x^2 + 2y^2 = 8$.

Answer: $y = \pm 2x + 6$.

Problem 201. Determine if the straight lines given by the following equations are tangential to the ellipse $x^2 + 4y^2 = 20$. If tangential, find the coordinates of the tangency point:

$$1)\ x + 4y - 10 = 0; \quad 2)\ 2x - y + 3 = 0; \quad 3)\ x + y - 5 = 0.$$

Answer: 1) is tangential at the point (2;2); 2) is not tangential; 3) is tangential at the point (4;1).

Problem 202. Write the equation of the tangent of the ellipse $x^2 + 4y^2 = 20$ perpendicular to the straight line $2x - 2y - 13 = 0$.

Answer: $x + y - 5 = 0$, $x + y + 5 = 0$.

Problem 203. Knowing that the line $A(-3;2)$ is on the ellipse, the straight line $4x - 6y - 25 = 0$ is tangential to this ellipse; write the equation of this ellipse.

Answer: $\frac{x^2}{25} + \frac{4y^2}{25} = 1, \frac{16x^2}{225} + \frac{9y^2}{100} = 1.$

Problem 204. Find the semi-axis, coordinates of focal points, eccentricity and asymptotes of the hyperbola given by the equation $\frac{x^2}{16} - \frac{y^2}{9} = 1$.

Answer: $a = 4$, $b = 3$, $F_1(-5;0)$, $F_2(5;0)$, $e = \frac{5}{4}$, $y = \pm\frac{3}{4}x$.

Problem 205. Write the canonical equation of the hyperbola by the given data:

1) The distance between the vertices equals 10, the distance between the focal points equals 12;
2) The length of the real axis of the hyperbola equals 1. The point (1;3) is on the hyperbola.

Answer: 1) $\frac{x^2}{25} - \frac{y^2}{11} = 1$; 2) $\frac{x^2}{\frac{1}{4}} - \frac{y^2}{3} = 1$.

Problem 206. On the hyperbola $\frac{x^2}{9} - \frac{y^2}{16} = 1$ find a point that the distance from this point to the focal point F_1 be equal to 7.

Answer: $\left(-6;\pm 4\sqrt{3}\right)$.

Problem 207. Write the equation of the straight line tangential to the hyperbola $\frac{x}{\frac{1}{4}}-\frac{y^2}{3}=1$ at the point (1;3).

Answer: $4x - y - 1 = 0$.

Problem 208. Write the equation of the tangent of the hyperbola $\frac{x^2}{20}-\frac{y^2}{5}=1$ perpendicular to the straight line $4x + 3y - 7 = 0$.

Answer: $3x - 4y - 10 = 0$, $3x - 4y + 10 = 0$.

Problem 209. Write the equation of the tangent drawn to the hyperbola $x^2 - y^2 = 16$ from the point $A(-1;-7)$.

Answer: $5x - 3y - 16 = 0$, $13x + 5y + 48 = 0$.

Problem 210. Write the equation of the hyperbola drawn to the straight lines $x = 1$ and $5x - 2y + 3 = 0$.

Answer: $x^2 - \frac{y^2}{4} = 1$.

Problem 211. Find canonic equation of the parabola by the following data:

1) The point (5; – 5) is on the parabola;
2) The distance from the focal point to the directrix equals 12;
3) $F(0;1)$ is a focal point, the parabola is symmetric with respect to the axis OY and is tangential to the axis OX.

Answer: 1) $y^2 = 5x$; 2) $y^2 = 24x$; 3) $x^2 = 4y$.

Problem 212. Write the equation of the straight line tangential to it at the point $M_0(x_0, y_0)$ on the parabola:

1) $y^2 = 5x$, $M_0(0{,}8;2)$;

2) $y^2 = 24x$, $M_0\left(\frac{1}{6};-2\right)$.

Answer: 1) $5x - 4y + 4 = 0$; 2) 2) $6x + y + 1 = 0$.

Problem 213. Write the equation of the tangent of the parabola $y^2 = 8x$ parallel to the straight line $2x + 2y - 3 = 0$.

Answer: $x + y + 2 = 0$.

Problem 214. Write the equation of tangent of the parabola $x^2 = 16y$ parallel to the straight line $2x + 4y + 7 = 0$.

Answer: $2x - y - 16 = 0$.

Problem 215. Write the equation of the tangent drawn to the parabola $y^2 = 36x$ from the point $A(2;9)$.

Answer: $3x - y + 3 = 0,\ 3x - 2y + 12 = 0$

Problem 216*. Determine the kind of the second-order curve whose equation is given in polar coordinates, and write its canonical equation.

$$1)\ r = \frac{144}{13 - 5\cos\varphi};\quad 2)\ r = \frac{18}{4 - 5\cos\varphi};\quad 3)\ r = \frac{2}{1 - \cos\varphi}.$$

Answer: $1)\ \frac{x^2}{169} + \frac{y^2}{144} = 1;\quad 2)\ \frac{x^2}{64} - \frac{y^2}{36} = 1;\quad 3)\ y^2 = 4x.$

Problem 217*. Accept the right focal point as a pole, the focal axis as a polar axis and write the following equations in polar coordinates:

$$1)\ \frac{x^2}{100} + \frac{y^2}{64} = 1;\quad 2)\ \frac{x^2}{144} - \frac{y^2}{25} = 1;\quad 3)\ y^2 = 8x.$$

Answer: $1)\ r = \frac{32}{5 - 3\cos\phi};\quad 2)\ r = \frac{25}{12 - 13\cos\varphi};\quad 3)\ r = \frac{4}{1 - \cos\varphi}.$

KEYWORDS

- **matrix**
- **elements**
- **columns**
- **determinant**
- **equipotential**

CHAPTER 2

INTRODUCTION TO MATHEMATICAL ANALYSIS

CONTENTS

ABSTRACT

In this chapter, we give theoretical materials about one-variable function, its domain of definition, one-variable function, its domain of definition, set of values, even and odd functions, periodic functions, numerical sequences, their limits, limit of a function, infinitely decreasing functions, continuous functions, discontinuity points, and 134 problems.

2.1 A FUNCTION OF ONE VARIABLE. DOMAIN OF DEFINITION OF A FUNCTION, SET OF VALUES. EVEN AND ODD FUNCTIONS. PERIODIC FUNCTIONS

Let X and Y be two sets of arbitrary real numbers.

Definition. If it is known the rule that associates to each value of x from X a unique value of y from Y, it is said that a function is given in the set X. X is called domain of definition of this function. The function is denoted as

$$y = f(x), y = \phi(x), y = g(x), \ldots, \text{ etc.}$$

x is called independent variable or argument, y the function of x.

In notation $y = f(x)$, the f is the characteristics of the function and shows the rule for finding appropriate value of y from Y to the given value of x from the X.

The subset of Y in the form $\{y | y = f(x), x \in X\}$ is said to be the set of values of the function $y = f(x)$ determined in X.

Function may be represented by different methods. The most used methods are analytic, tabular, and graphic.

When a function is represented by the analytic method, the dependence of y on x is shown by one or several formulas. For example, the functions

$$y = 2x + 3, \; y = \begin{cases} x^2, & x \le 0, \\ \sin x, & x > 0, \end{cases}$$

$$y = \begin{cases} -1, & x < 0, \\ 0, & x = 0, \\ 1, & x > 0 \end{cases}$$

were represented by the analytic method. In mathematics for the last function a special sign was accepted:

$$signx = \begin{cases} -1, & x < 0, \\ 0, & x = 0, \\ 1, & x > 0. \end{cases}$$

When a function is represented by the tabular method, the values $y_1 = f(x_1), y_2 = f(x_2), ..., y_n = f(x_n)$ of y appropriate to several $x_1, x_2, \dots, x_n$ values of the argument x are given in the form of a table.

In the graphic method, in the coordinate system on a plane such a curve is given that any straight line perpendicularly drawn to the abscissa axis intersects this curve only at one point. The y_0 ordinate of the point that intersects the curve perpendicularly drawn to the abscissa axis at the point with the abscissa x_0 is the value of this function appropriate to the number $x = x_0$: $y_0 = f(x_0)$

When a function is represented by the analytic method, the dependence of y on x may be given explicitly not by the formula, $y = f(x)$, the dependence between x and y may be represented in the form $F(x, y) = 0$. In this case, it is said that $y = y(x)$ is given implicitly by the equation $F(x, y) = 0$. For example, the function $y = 3x + 2$ may be represented implicitly by the equation $3x - y + 2 = 0$. In giving a function in the analytic form, the variables x and y may be represented in the form of functions of another variable t by the formula

$$x = \varphi(t),\ y = \psi(t),\ t \in T.$$

In this case it is said that a function is represented in the parametric form. t is called a parameter.

A function given by the formula $y = f(x),\ x \in X$ in the parametric form is written as $x = t,\ y = f(t), t \in X$. Sometimes, it is easy to represent explicitly the function given in the parametric form by one formula. For example, let us represent explicitly the function given in the parametric form by one formula

$$x = t + 1,\ \ y = 2t - 3,\ t \in (-\infty, +\infty).$$

For that we must find t ($t = x - 1$) from the expression of x and write in the expression of y:

$$y = 2t - 3 = 2(x-1) - 3 = 2x - 5 \Rightarrow y = 2x - 5, x \in (-\infty, +\infty).$$

Representation of a function may be described by words. For example, the Dirichlet function is expressed by words in the following form:

$$D(x) = \begin{cases} 1, & \text{when } x \text{ is a rational number,} \\ 0, & \text{when } x \text{ is an irrational number.} \end{cases}$$

The function that represents the greatest number not exceeding x and called "entire part of x" or "the greatest integer x" may be denoted by

$$y = [x].$$

For example, $[1,2] = 1$; $[-1,5] = -2$.

If the function $x = \varphi(t)$ is defined in the set T and its set of values enters into the domain of definition of the function $y = f(x)$, then the function $y = f(\varphi(t))$ is called a complex function of variable t determined on the set T. For example, if for the function $y = \sin(t^4 + 1)$ we take $y = \sin x$, $x = t^4 + 1$) we can consider the given function as a complex function of t.

Definition. If the function $y = f(x)$ determined in the set X satisfies the conditions

$$f(x_1) \le f(x_2) (f(x_1) < f(x_2), f(x_1) \ge f(x_2), f(x_1) > f(x_2))$$

for arbitrary numbers x_1, x_2 from X determining the condition x_1, $< x_2$, then $f(x)$ is said to be a non-decreasing (increasing, non-increasing, and decreasing) function. A function that satisfies any of these conditions is called monotone in the set X. Increasing and decreasing functions are said to be strongly monotone functions.

One and the same function may be increasing in one part of the domain of definition, and in another part may be decreasing. For example, the function $y = x^2$ is decreasing on the interval $(-\infty; 0)$, is increasing on the interval $(0; +\infty)$.

If for the function $y = f(x)$ determined in the set X there exists a number $M(m)$ such that for all $x \in X$ $f(x) \le M$ $(f(x) \ge m)$, then $f(x)$ is called a function bounded above (below) in the set X. A function bounded both above and below in the set X is said to be a bounded function in this set. For every bounded function $f(x)$ in the set X it is possible to find a number $A > 0$ such that for all $x \in X$, $|f(x)| < A$.

Let y be the set of values of the function $y = f(x)$ determined in the set X. If there is a rule that associates to every value of y from Y a unique number x from X satisfying the equality $f(x) = y$, then this rule represents a function determined in Y and this function is called the inverse function of $y = f(x)$ and is denoted by $x = f^{-1}(y)$. For every $y \in Y$ the relation $f\left(f^{-1}(y)\right) = y$ and for any $x \in X$ the relation $f^{-1}(f(x)) = x$ are true. These two relations are accepted as definition of relative inverse functions. Sometimes the inverse function of the function $y = f(x)$ is denoted as $y = \varphi(x)$. This time it should be taken into an account that the domain of definition of the argument x of the function $y = \varphi(x)$ is the set Y (the set of values of $y = f(x)$). The graphs of the functions $y = f(x)$ and $y = \varphi(x)$ are symmetric on the same system of coordinates with respect to bisectrix of the first coordinate quarter. The following statement is true.

Theorem. The strongly monotone function $y = f(x)$ determined in the set X has an inverse function in the set Y, and this inverse function is strongly monotone in the set Y.

If for a function $f(x)$ determined in the set X there exists a number $T \neq 0$ such that for all $x \in X$, $f(x + T) = f(x)$, $(x + T \in X)$, then $f(x)$ is called a periodical function in the set X and T is called a period of this function. If the number T is the period of $f(x)$, and all the points in the form of $x + nT$ enter into the domain of definition of this function, then the numbers $2T, 3T, 4T, ..., nT$ are periods of $f(x)$. Sometimes under a period of $f(x)$ one understands the least positive number satisfying the equality $f(x + T) = f(x)$.

The intervals in the form

$$(-\infty, +\infty),\ (-a, a),\ [-a, a],\ (a > 0)$$

are called symmetric sets with respect to the origin of coordinates.

Definition. For the function $f(x)$ determined in the set X symmetric with respect to the origin of coordinates, if

$f(-x) = f(x)$, then $f(x)$ is called an even function and
if $f(-x) = -f(x)$, then $f(x)$ is called an odd function.

The graphs of even functions are symmetric with respect to ordinate axis, the graphs of odd functions with respect to origin of coordinates.

If for the function $f(x)$ given in the symmetric set we take

$$\varphi(x) = \frac{1}{2}\left(f(x) + f(-x)\right),$$
$$\psi(x) = \frac{1}{2}\left(f(x) - f(-x)\right),$$

then, $\varphi(x)$ is an even, $\psi(x)$ is an odd function. Therefore, every $f(x)$ function determined in symmetric set may be uniquely represented in the form of the sum of one even and one odd function:

$$f(x) = \varphi(x) + \psi(x).$$

There are such functions that are neither even nor odd.

Problem to be solved in auditorium

Problem 218. In Figure 2.1 accordance between the sets A and B is given by axes. Determine which of these accordances is a function.

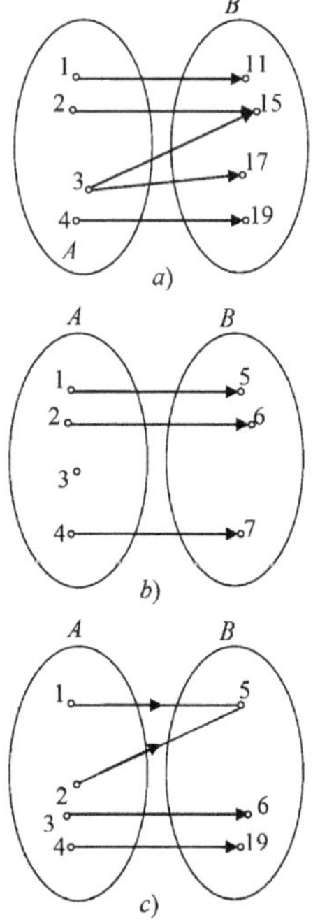

FIGURE 2.1

Answer: *a)* is not a function;
b) is not a function;
c) is a function.

Problem 219. Accordance R between two sets is given by the pairs in the form $(x; y)$. In the pair $(x; y)x$ represents the element of the first set, y represents the element assigned to x. Is this accordance a function? If it is a function, represent its domain of definition and the set of values:

$$1)\ R=\{(1;2),(3;4),(5;6),(7;8),(9;10)\};$$
$$2)\ R=\{(1;2),(1;3),(1;4),(1;5)\};$$
$$3)\ R=\{(1;2),(2;2),(3;2),(4;2),(5;2)\};$$
$$4)\ R=\{(1;1),(2;2),(3;3),(4;4),(5;5)\};$$
$$5)\ R=\{(1;0),(-1;0),(2;2),(-2;2),(-2;-2)\}.$$

Answer: 1) is a function, the domain of definition $\{1;3;5;7;9\}$ the set of values $\{2;4;6;8;10\}$;
2) is not a function;
3) is a function, the domain of definition $\{1;2;3;4;5\}$, the set of values $\{2\}$;
4) is a function, the domain of definition $\{1;2;3;4;5\}$, the set of values $\{1;2;3;4;5\}$;
5) is not a function.

Problem 220. Find the domain of definition of the following functions:

$$1)\ y=\frac{x}{x+1};\ 2)\ y=\frac{x^3-1}{x^2-6x+8};\ 3)\ y=\frac{1}{x+|x|};$$
$$4)\ y=\frac{|x+2|+1-2x-2x^2}{|2x+2|-1};\ 5)\ y=\sqrt[3]{x^2-1};$$
$$6)\ y=\sqrt[4]{2-x};\ 7)\ y=\sqrt{2^x-3^x};\ 8)\ y=\frac{1}{16^{x^2}-2}.$$

Answer: $1)(-\infty;-1)\cup(-1;+\infty); 2)(-\infty;2)\cup(2;4)\cup(4;+\infty);$

$3)(0;+\infty); 4)(-\infty;-1\frac{1}{2})\cup(-1\frac{1}{2};-\frac{1}{2})\cup(\frac{1}{2};+\infty);$

$5)(-\infty;+\infty); 6)(-\infty;2]; 7)(-\infty;0];$

$8)(-\infty;0)\cup(0;0,25)\cup(0,25;+\infty).$

Problem 221. Find the set of values of the functions:

$$1)\ f(x)=2x-5, x\in[-2;2]; 2)\ f(x)=x+signx, x\in(-\infty;+\infty);$$

$$3)\ f(x)=-2x^2+x+1, x\in(-\infty;+\infty); 4)\ f(x)=x+\frac{1}{x}, x\in(0;+\infty);$$

$$5)\ f(x)=\sqrt{x^2+1}, x\in(-\infty;+\infty); 6)\ f(x)=\sqrt{x(4-x)}, x\in[0;4];$$

$$7)\ f(x)=4^x-2^x+1, x\in(-\infty;+\infty); 8)\ f(x)=\log_3 x+\log_x 3.$$

Answer: $1)[-9;-1]; 2)(-\infty;-1)\cup\{0\}\cup(1;+\infty);\ 3)\left(-\infty;1\frac{1}{8}\right];$

$4)[2;+\infty); 5)[1;+\infty); 6)[0;2];\ 7)\left[\frac{3}{4};+\infty\right); 8)(-\infty;-2)\cup[2;+\infty).$

Problem 222. For $f(x)=\begin{cases}1+x, & -\infty<x\le 0,\\ 2^x, & 0<x<+\infty\end{cases}$ find $f(-2), f(-1), f(0), f(1), f(2)$.

Answer: $f(-2)=-1, f(-1)=0, f(0)=1, f(1)=2, f(2)=4.$

Problem 223. Knowing that $\alpha_n=\varphi(n)$ denotes a function that assigns the natural number n to the n-th decimal point of the number $\sqrt{2}$, find $\varphi(1)$, $\varphi(2)$, $\varphi(3)$, $\varphi(4)$:

Solution. It is known that $\sqrt{2}=1,4142\ldots$. Therefore find $\varphi(1)=4$, $\varphi(2)=1$, $\varphi(3)=4$, $\varphi(4)=2$.

Problem 224. For $f(x)=\frac{1-x}{1+x}$

find $f(0), f(-x), f(x+1), f(x)+1, f\left(\frac{1}{x}\right), \frac{1}{f(x)}$.

Answer: $f(0)=1, f(-x)=\frac{1+x}{1-x}, f(x+1)=\frac{-x}{2+x},$

$$f(x)+1=\frac{2}{1+x}, f\left(\frac{1}{x}\right)=\frac{x-1}{x+1}, \frac{1}{f(x)}=\frac{1+x}{1-x}.$$

Problem 225. The function $f(x)$ was determined on the interval $[0;1]$. Find the domain of definition of the following functions:

$$1)\ f\left(3x^2\right);\ 2)\ f(x-5);\ 3)\ f(tgx).$$

Solution of 1): The given function is a complex function of the argument x. If we denote $u = 3x^2$, as by the condition the function $f(x)$ is determined in the interval $[0;1]$, we can write

$$0 \le u \le 1 \Rightarrow 0 \le 3x^2 \le 1 \Rightarrow 0 \le x^2 \le \frac{1}{3} \Rightarrow -\frac{1}{\sqrt{3}} \le x \le \frac{1}{\sqrt{3}}.$$

Thus, under the given condition the domain of definition of the function $f(3x^2)$ is the interval $\left[-\frac{1}{\sqrt{3}}; \frac{1}{\sqrt{3}}\right]$.

Answer: $2)[5;6];\ 3)\ k\pi \le x \le \frac{\pi}{4}+k\pi, (k=0,\pm1,\pm2,\ldots)$.

Problem 226. Find a function in the form $f(x)=ax^2+bx+c$ such that $f(0)=5, f(-1)=10; f(1)=6$.

Answer: $f(x)=3x^2-2x+5$.

Problem 227. Show that the function $y=\frac{1}{x^2}, x\in(-\infty;0)\cup(0;+\infty)$ is bounded lower but is not bounded above.

Solution. As this function gets only positive values, it is clear that this function is bounded below: $y > 0$.

Show that this function is not bounded above. If this function is not bounded above, this means that for any $C > 0$ we can find a value of x such that, $f(x) > C$. Indeed, if we take $x=\frac{1}{2\sqrt{C}}$, then

$$f\left(\frac{1}{2\sqrt{C}}\right)=\frac{1}{\left(\frac{1}{2\sqrt{C}}\right)^2}=4C>C.$$

Thus, the given function is not bounded above.

Problem 228. Show that for $a > 0$ the function $y = ax^2 + bx + c$ is not bounded above, for $a < 0$ is not bounded below.

Problem 229. Show that the following functions are bounded in the given sets:

$$1)\ y = \frac{x}{x^2+1}, x \in (-\infty; +\infty);$$

$$2)\ y = x^2 - x - 1, x \in [-1; 5];$$

$$3)\ y = \frac{3x^2 + 6x + 10}{\sqrt{0,1x^4 + 1}}.$$

Guideline: In 1) use the relation $|x| \le \frac{x^2+1}{2}$.

Problem 230. Find the largest and the least values of the represented functions in the set X:

$$1)\ f(x) = x^2 - 4x - 5, X = [0; 5];$$
$$2)\ f(x) = -x^2 + 5x - 6, X = [0; 4].$$

Solution of 2): represent the given function in the following form:

$$f(x) = -\left(x^2 - 5x + 6\right) = -\left((x-2.5)^2 - 0.25\right) = 0.25 - (x-2.5)^2 .$$

From the last expression of the function it is seen that this function gets its largest value for $x = 2.5 \quad \left(2.5 \in [0;\ 4]\right)$:

$$\max \quad f = f(2.5) = 0.25.$$

This function gets its least value on $X = [0;4]$ for $x = 0$:

$$\min f(0) = 0.25 - 6.25 = -6.$$

Answer: $1) \max f = f(5) = 0; \quad \min f = f(2) = -9.$

Problem 231. Find the increase and decrease intervals of the following functions:

$$1)\ f(x) = x^2 - 4x - 5; 2)\ f(x) = -x^2 + 5x - 6;$$
$$3)\ f(x) = \sin x + \cos x.$$

Solution of 2): Write the given function in the form

$$f(x) = -(x-2.5)^2 + 0.25 .$$

From the last formula we see that geometrically this function is described by a parabola with the vertex at the point (2.5; 0.25) and with branches directed downward. Therefore, it is clear that for $x < 2.5$, that is, on the interval $(-\infty;\ 2.5)$ the following function increases for $x < 2,5$, that is, on the interval $(2.5;\ +\infty)$ the function decreases.

Answer: 1) 1) on the interval $(-\infty; 2)$ it decreases; on the interval $(2; +\infty)$ it increases; 3) on the interval $\left[\frac{\pi}{4} + 2n\pi; \frac{\pi}{4} + (2n+1)\pi\right]$ it decreases on the interval $\left[\frac{\pi}{4} + (2n-1)\pi; \frac{\pi}{4} + 2n\pi\right]$ $(n = 0, \pm 1, \dots)$ it increases.

Guideline: Represent the given function in the form $f(x) = \sqrt{2}\cos\left(x - \frac{\pi}{4}\right)$.

Problem 232. Define which of the following functions is odd, is even and neither odd nor even:

1) $f(x) = x^4 + 5x^2,\ x \in (-\infty; +\infty)$;

2) $f(x) = \dfrac{x^3}{x^2+1},\ x \in (-\infty; +\infty)$;

3) $f(x) = \sin x - \cos x,\ x \in (-\infty; +\infty)$;

4) $f(x) = \dfrac{x^6}{x^2+1},\ x \in (-\infty; 1]$;

5) $f(x) = |x+1| + |x-1|,\ x \in (-\infty; +\infty)$;

6) $f(x) = \lg\dfrac{1+x}{1-x},\ x \in (-1; 1)$;

7) $f(x) = \begin{cases} x^4,\ x > 0, \\ x^2,\ x \le 0; \end{cases}$

8) $f(x) = \lg\left(x + \sqrt{1+x^2}\right)$.

Solution of 8): At first we note that this function was determined on the real axis. We can easily see that for the given function $f(x) + f(-x) = 0$. Indeed,

$$f(x)+f(-x)=\lg\left(x+\sqrt{1+x^2}\right)+\lg\left(-x+\sqrt{1+(-x)^2}\right)=$$
$$=\lg\left(\sqrt{1+x^2}+x\right)+\lg\left(\sqrt{1+x^2}-x\right)=\lg\left(1+x^2-x^2\right)=\lg 1=0.$$

From the relation $f(x)+f(-x)=0$ we get $f(-x)=-f(x)$. So, the given function is odd.

Answer: 1) is even; 2) is odd; 3) is neither odd nor even; 4) is neither odd nor even; 5) is even; 6) is odd; 7) is neither odd nor even.

Problem 233. Prove that the product of two odd functions and two even functions is an even function. The product of an odd function and even function is an odd function.

Problem 234. Prove that if the function $f(x)$ is of period T, then for $a < 0$, the function $f(ax+b)$ is a periodic function of period T/a.

Solution. At first, show that if T is the period of $f(x)$, the number T/a is one of the periods of the function $f(ax + b)$, $a > 0$. Indeed,

$$f\left(a\left(x+\frac{T}{a}\right)+b\right)=f\left(ax+T+b\right)=f\left(\left(ax+b\right)+T\right)=f(ax+b).$$

Now show that no equation less than T/a may be a period of the function $f(ax + b)$.

Suppose that the number $T_1 > 0$ is an arbitrary period of the function $f(ax + b)$, that is,

$$f\left(a\left(x+T_1\right)+b\right)=f(ax+b). \tag{2.1}$$

From the domain of definition of $f(x)$ take any number x and denote $x'=\frac{x-b}{a}$. Then we can write:

$$f(ax'+b)=f\left(a\cdot\frac{x-b}{a}+b\right)=f(x-b+b)=f(x).$$ So, we get

$$f(x)=f(ax'+b) \tag{2.2}$$

In eq 2.1 instead of x we write x':

$$f(ax'+b)=f\left(a\left(x'+T_1\right)+b\right). \tag{2.3}$$

Based on equalities (2.2) and (2.3), we can write:

$$f(x) = f(ax'+b) = f\left(a(x'+T_1)+b\right) = f\left(ax'+aT_1+b\right) =$$
$$= f\left(a\cdot\frac{x-b}{a}+b+aT_1\right) = f\left(x+aT_1\right) \Rightarrow f(x) = f\left(x+aT_1\right).$$

The last relation shows that the number aT_1 is one of the periods of $f(x)$. According to the condition, as T is the least positive period of $f(x)$, there should be $aT_1 \geq T$. Hence we get $T_1 \geq T/a$, that is, when T is the least positive period of $f(x)$, the least positive period of the function $f(ax + b)$ is the number $T/a\ (a > 0)$.

Problem 235. Find the least positive periods of the functions:

1) $f(x) = \sin 5x$; 2) $f(x) = ctg\frac{x}{2}$;

3) $f(x) = \sin 2\pi x$; 4) $f(x) = \sin 3x + \cos 2x$;

5) $f(x) = \cos 2x \cdot \cos 6x$.

Solution of 5): At first represent $f(x)$ as follows:

$$f(x) = \cos 2x \cdot \cos 6x = \frac{1}{2}(\cos 8x + \cos 4x).$$

The least positive period of the function $f_1(x) = \frac{1}{2}\cos 8x$ is $\frac{2\pi}{8} = \frac{\pi}{4} = 45°$, the least positive period of the function $f_2(x) = \frac{1}{2}\cos 4x$ is $\frac{2\pi}{4} = 90°$ (see problem 234). Then it is clear that the period of the function $f(x)$ that is the sum of these functions, at the same time is the least positive number being the period both of the functions $f_1(x)$ and $f_2(x)$. As the periods of $f_1(x)$ is in the form of total product of 45°, the periods of $f_2(x)$ in the form of total product of 90°, the period of the function $f(x)$ must be the least bisectrix of 45° and 90°. So, the least positive period of the given function is $90^\circ = \pi/2$.

Answer: 1) $\frac{2\pi}{5}$; 2) 2π; 3) 1; 4) 2π.

Problem 236. Prove that

$$D(x)=\begin{cases}1, & \text{when } x \text{ is a rational number,}\\ 0, & \text{when } x \text{ is an irrational number.}\end{cases}$$

The Dirichlet function is periodic and every nonzero rational number is its period, no irrational number is its period.

Solution. Let r be a rational number. Then as for arbitrary rational number x, $x + r$ is also a rational number, then

$$D(x+r) = D(x) = 1 \cdot$$

When x is an irrational number, as $x + r$ is an irrational number, then $D(x+r) = D(x) = 0$.

So, as in all numbers of x, for every rational number r $D(x + r) = D(x)$, any rational number is the period of the function $D(x)$.

Now take an arbitrary irrational number γ. At rational values of x, $x + \gamma$ will be irrational. Then by the rule of definition of the function $D(x)$,

$$D(x) = 1 \quad D(x + \gamma) = 0.$$

$D(x) \neq D(x + \gamma)$ means that no irrational number is a period of $D(x)$.

Problem 237. Prove that the function $f(x) = \cos x^2$ is not periodic.

Solution. Let us assume the converse. Assume that the function $f(x)$ is a periodic function of period T. Then at any value of x the equality $\cos(x+T)^2 = \cos x^2$ should be valid. Hence it follows that at any value of x the identity

$$x^2 + 2Tx + T^2 \pm x^2 \equiv 2k\pi,\ k = 0,\ \pm 1,\ \pm 2,\ldots$$

should be valid. This is impossible.

Problem 238. Show if the following functions have inverse functions and find these inverse functions:

$$1)\, y = f(x) = -3x + 5; 2)\, y = f(x) = x^2 - x + 1,\ x \in \left[\frac{1}{2}; +\infty\right);$$

$$3)\, y = 5^{\lg x},\ (x > 0).$$

Solution of 2): Represent the given function in the form $f(x) = x^2 - x + 1 = \left(x - \frac{1}{2}\right)^2 + \frac{3}{4}$. It is clear from the last expression of the function that this function is increasing on the interval $X = \left[\frac{1}{2}; +\infty\right)$. The given function associates two values of x from the set X with the values of y from the set $Y = \left[\frac{3}{4}; +\infty\right)$. According to the theorem on the existence of the inverse function, the given function has an inverse function in the set Y. This function is found by solving the following equation with respect to x:

$$x^2 - x + 1 = y \Rightarrow x^2 - x + (1 - y) = 0 \Rightarrow x = \frac{1}{2} + \sqrt{y - \frac{3}{4}}.$$

Thus, the given function has an inverse function and this function is represented by the formula,

$$x = f^{-1}(y) = \frac{1}{2} + \sqrt{y - \frac{3}{4}}, \ y \in \left[\frac{3}{4}; +\infty\right).$$

Answer: $1)\, x = -\frac{1}{3}y + \frac{5}{3};\ 3)\, x = y^{\frac{1}{\lg 5}}, (y > 0).$

Problem 239. Construct the graphs of the following functions:

$$1)\, sign\, x = \begin{cases} 1, \ x > 0, \\ 0, \ x = 0, \\ -1, \ x < 0; \end{cases}$$

$$2)\, y = |x - 1|, \ x \in (-\infty; +\infty);$$

$$3)\, y = x - 4 + |x - 2|, \ x \in [-2; 5].$$

Problem 240. For $f(x) = 2x + 1$ find the graphs of the functions $f(x)$, $-f(x)$, $f(-x)$.

Home tasks

Problem 241. The correspondence R between two sets is given by the pairs in the form $(x; y)$. In the pair $(x; y)$, x represents the elements of the first set, y represents the elements assigned to x. Is this congruence a function? If it is a function, represent its domain of definition X and the set of values Y:

$$1)\, R=\{(-3;9),(-2;4),(-1;1),(0;0),(1;1),(2;4),(3;9)\};$$

$$2)\, R=\{(9;-3),(4;-2),(1;-1),(0;0),(1;1),(4;2),(9;3)\};$$

$$3)\, R=\{(-3;-1),(-2;-1),(-1;-1),(0;0),(1;1),(2;1),(3;1)\};$$

$$4)\, R=\{(\sqrt{2};0),(\sqrt{3};0),(2;1),(3;1)\};$$

$$5)\, R=\{(-1;-1),(-1;0),(1;1),(1;3)\};$$

$$6)\, R=\{(-3;3),(-2;2),(-1;1),(0;0),(1;1),(2;2),(3;3)\};$$

$$7)\, R=\{(3;-3),(2;-2),(1;-1),(0;0),(1;1),(2;2),(3;3)\}.$$

Answer: 1) is a function, $X=\{-3;-2;-1;0;1;2;3\}$, $Y=\{0;1;4;9\}$; 2) is not a function; 3) is a function, $X=\{-3;-2;-1;0;1;2;3\}$, $Y=\{-1;0;1\}$; 4) is a function, $X=\{\sqrt{2},\sqrt{3},2;3\}$, $Y=\{0;1\}$; 5) is not a function; 6) is a function, $X=\{-3;-2;-1;0;1;2;3\}$, $Y=\{0;1;2;3\}$; 7) is not a function.

Problem 242. Find the domain of definition of the following functions:

$$1)\, y=\frac{x+1}{x^2-1};2)\, y=\frac{(x+2)^2}{x^3-4x};3)\, y=\frac{x^2}{2|x|-3};$$

$$4)\, y=\sqrt{-x^2};5)\, y=\sqrt{2-x-x^2}.$$

Answer: $1)(-\infty;-1)\cup(-1;1)\cup(1;+\infty);$

$$2)(-\infty;-2)\cup(-2;0)\cup(0;2)\cup(2;+\infty);$$

$$3)\left(-\infty;-1\frac{1}{2}\right)\cup\left(-1\frac{1}{2};1\frac{1}{2}\right)\cup\left(1\frac{1}{2};+\infty\right);$$

$$4)\{0\};\quad 5)[-2;1].$$

Problem 243. When the function f_1 and f_2 are given by the following formulas, find the domain of definition of the functions f_1, f_2 and f_1+f_2:

$$1)\, f_1(x)=\sqrt[4]{3-x},\, f_2(x)=\sqrt{x+1};$$

$$2)\, f_1(x)=\sqrt{1-x^2},\, f_2(x)=\sqrt[3]{\frac{x}{2x-1}}.$$

Answer: 1) $X_{f_1} = (-\infty;3], X_{f_2} = [-1;+\infty), X_{f_1+f_2} = [-1;3]$;

$$2)\, X_{f_1} = [-1;1], X_{f_2} = \left(-\infty;\frac{1}{2}\right) \cup \left(\frac{1}{2};+\infty\right),$$

$$X_{f_1+f_2} = \left[-1;\frac{1}{2}\right) \cup \left(\frac{1}{2};1\right].$$

Problem 244. Find the set of values of the functions:

1) $f(x) = |x-1|,\ x \in [0;5]$;

2) $f(x) = x^2 + 2x - 3, x \in (-\infty;+\infty)$;

3) $f(x) = 5 - 12x - 2x^2, x \in [-4;1]$;

4) $f(x) = \dfrac{x^2+4}{x}, x \in (-\infty;0)$.

Answer: 1) $[0;4]$; 2) $[-4;+\infty]$; 3) $[-9;23]$; 4) $(-\infty;-4]$.

Problem 245. The function $f(x)$ was determined on the interval $[0;1]$. Find the domain of definition of the following functions:

$$1)\, f(\sin x);\quad 2)\, f(2x+3).$$

Answer: 1) $x \in [2k\pi;(2k+1)\pi], (k = 0, \pm 1, \pm 2, \ldots)$;

$$2)\, x \in \left[-\frac{3}{2};-1\right].$$

Problem 246. At the points satisfying the condition $4/x + x = 5$ find the values of the functions $f(x) = \dfrac{16}{x^2} + x^2$.

Answer: 17.

Guideline: Use the relation $\dfrac{16}{x^2} + x^2 = \left(\dfrac{4}{x} + x\right)^2 - 8$.

Problem 247. The function $f(x)$ was given by the formula

$$f(x) = \begin{cases} 2x^3 + 1, x \le 2, \\ \dfrac{1}{x-2}, 2 < x \le 3, \\ 2x - 5, x > 3. \end{cases}$$

Find $f\left(\sqrt{2}\right)$, $f\left(\sqrt{8}\right)$, $f\left(\sqrt{\log_2 1024}\right)$.

Answer: $f\left(\sqrt{2}\right)=4\sqrt{2}+1$, $f\left(\sqrt{8}\right)=\frac{\sqrt{2}+1}{2}$,

$f\left(\sqrt{\log_2 1024}\right)=2\sqrt{10}-5$.

Problem 248. Knowing $f(x+1)=x^2-3x+2$, find $f(x)$.

Answer: $f(x)=x^2-5x+6$.

Problem 249. Show that the following functions are bounded:

$$1)\ y=\frac{1}{x-10},\ x\in[0;5];$$

$$2)\ y=\frac{x^2}{1+x^4},\ x\in(-\infty;+\infty).$$

Guideline: in 2) use that, $\left(1-x^2\right)^2\geq 0\Rightarrow 1+x^4\geq 2x^2$.

Problem 250. Show that on the interval $x\in(-\infty;0)\cup(0;+\infty)$ the function $y=1/x$ is bounded neither above nor lower.

Problem 251. Show that on the interval the function $y=\frac{1}{\sqrt{1-x^2}}$ is bounded lower, is not bounded above $x\in(-1;1)$.

Problem 252. Show that for $x\in(-\infty;+\infty)$ the function $f(x)=x^2-3x$ has no greatest value, it has the least value. Find the least value of this function.

Answer: $\min f=-\frac{9}{4}$.

Problem 253. Show that for $x\in(-\infty;+\infty)$ the function $f(x)=4x-x^2-6$ has no least value, but has the greatest value. Find the greatest value of this function.

Answer: $\max f=-2$.

Problem 254. Find the least value of the function $f(x)=3^{\left(x^2-2\right)^3+8}$.

Answer: $\min f=3^{\circ}=1$.

Guideline: Represent the exponent of 3 in the form $x^2\left(\left(x^2-3\right)^2+3\right)$.

Problem 255. Among the rectangles with perimeter $2P>0$ find the lengths of the sides of one with the largest area.

Answer: A quadrate with the side $P/2$.

Guideline: Denote the length of one side of the sought-for rectangle by x, express the S area of the rectangle by P and x, and for finding the greatest value of $S(x)$ write it in the form $S(x)=\frac{P^2}{4}-\left(x-\frac{P}{2}\right)^2$.

Problem 256. Show that the function $f(x)=x^3+x$ increases on the whole axis.

Guideline: Take two real numbers satisfying the condition $x_1>x_2$ and formulate their difference in the form $f(x_1)-f(x_2)$

$$(x_1-x_2)\left(\left(x_1+\frac{1}{2}x_2\right)^2+\frac{3}{4}x_2^2+1\right).$$

Problem 257. Find the increase and decrease intervals of the function $y=tg\left(x+\frac{\pi}{3}\right)$.

Answer: Increases on the interval $\left(-\frac{5\pi}{6}+k\pi,\frac{\pi}{6}+k\pi\right)$, $(k=0,\pm1,\pm2,\ldots)$ of the number axis, decreases on all remaining intervals.

Problem 258. Determine which of the functions is odd or even and neither even nor odd:

1) $f(x)=\frac{1}{1-x^3}, x\in(-1;1)$;

2) $f(x)=\frac{2^x+1}{2^x-1}, x\in(-\infty;+\infty)$;

3) $f(x)=\lg\frac{x+3}{x-3}, x\in(-\infty;-3)\cup(3;+\infty)$;

4) $f(x)=|x|+2, x\in(-\infty;+\infty)$;

5) $f(x)=|x+2|, x\in(-\infty;+\infty)$;

6) $f(x)=2^x+2^{-x}$;

$7)\ f(x)=x^2\sqrt[3]{x}+2\sin x;$

$8)\ f(x)=\sqrt{1+x+x^2}-\sqrt{1-x+x^2};$

$9)\ f(x)=const.$

Answer: 1) is neither odd nor even; 2) is odd; 3) is odd; 4) is even; 5) is neither even nor odd; 6) is even; 7) is odd; 8) is odd; 9) is even.

Problem 259. Determine periodical functions among the following functions and find the least positive periods:

$1)\ f(x)=5\cos 3x;\ 2)\ f(x)=\cos^2 3x;$

$3)\ f(x)=x\sin x;\ 4)\ f(x)=\cos x+\sin\left(\sqrt{3}x\right);$

$5)\ f(x)=\sin x^2;\ 6)\ f(x)=tg\frac{x}{2}-2tg\frac{x}{3};$

$7)\ f(x)=tg3x+\cos 4x;\ 8)\ f(x)=\sin 2x+\sin^2 3x;$

$9)\ f(x)=\sin^4 x+\cos^4 x;\ 10)\ f(x)=tg\left(x+\sin x\right).$

Answer: 1) $T=\frac{2\pi}{3}$; 2) $T=\frac{\pi}{3}$; 3) is not periodic; 4) is not periodic; 5) is not periodic; 6) $T=6\pi$; 7) $T=\pi$; 8) $T=\pi$; 9) $T=\frac{\pi}{2}$; 10) $T=2\pi$.

Problem 260. Prove that the following functions are mutually inverse:

$$1)\ y=f(x)=\frac{x}{x+1},(x\neq -1),\ \ x=f^{-1}(y)=\frac{y}{1-y},(y\neq 1);$$

$$2)\ y=f(x)=\sqrt[3]{1-x^3},\ \ x=f^{-1}(y)=\sqrt[3]{1-y^3}.$$

Guideline: Show that both of the relations $f\left(f^{-1}(y)\right)=y$ and $f^{-1}\left(f(x)\right)=x$ are satisfied.

Problem 261. Determine the inverse functions among the following functions and represent their inverse functions by formula:

$1)\ y=2x-1;\ 2)\ y=|x|;\ 3)\ y=\frac{1}{x^3};$

$4)\ y=x^2+2x-3;\ 5)\ y=sign x;$

$6)\ y=x^2-1,\ x\in\left(-\infty;-\frac{1}{2}\right];$

$$7)\, y = x^2 - 1,\ x \in \left[\frac{1}{2};+\infty\right).$$

Answer: 1) $x = \frac{y+1}{2}$; 2) on the interval $(-\infty;+\infty)$ has no inverse function; 3) $x = y^{-\frac{1}{3}}$; 4) on the interval $(-\infty;+\infty)$ has no inverse function; 5) $(-\infty;+\infty)$ on the interval has no inverse function;

6) $x = -\sqrt{y+1}, y \in \left[-\frac{3}{4};+\infty\right)$; 7) $x = \sqrt{y+1}, y \in \left[-\frac{3}{4};+\infty\right)$.

Problem 262. Construct the graphs of the following functions:

$$1)\, f(x) = \begin{cases} -x^2, & x < 0, \\ 3x, & x \geq 0; \end{cases} \quad 2)\, f(x) = \begin{cases} 4 - x, & x < -1, \\ 5, & -1 \leq x < 0, \\ x^2 + 5, & x > 0; \end{cases}$$

$$3)\, f(x) = x^2 + x - |x|.$$

Problem 263. For $f(x) = x^2$, $\varphi(x) = 2^x$ find $f(\varphi(x))$ and $\varphi(f(x))$.

Answer: $f(\varphi(x)) = 2^{2x}$, $\varphi(f(x)) = 2^{x^2}$.

Problem 264. Determine if the points A and B with the known coordinates belong to the graph of function given by the parametric equations, $x = \varphi(t)$, $y = \psi(t)$:

$$1)\, x = t^2 - 1,\ y = t^3 - t,\ A(0;0), B(3;3);$$
$$2)\, x = \sin t + 1, y = \cos t - 1, A(0;-1), B(1,6;-0,2);$$
$$3)\, x = 2^t \sin t,\ y = 2^t \cos t, A(2;2), B\left(0;2^{\pi}\right).$$

Answer: 1) A belongs, B does not; 2) both A and B belong; 3) neither A nor B belong.

Problem 265. Find the notation in the ordinary form $y = f(x)$ of the following functions given in the implicit form and construct their graphs:

$$1)\, \frac{xy-1}{y-x} = 0;\ 2)\, \frac{y-2x^2}{y-8} = 0.$$

Answer: 1) $y=\frac{1}{x}, x\neq 0, x\neq \pm 1$; 2) $y=2x^2, x\neq \pm 2$.

2.2 NUMERICAL SEQUENCES. LIMIT OF SEQUENCES

The function whose domain of definition is the set of all natural numbers N is called a number sequence. In other words, if there is a rule that associates every natural number n with the number x_n, it is said that the number sequence $x_1, x_2, \ldots, x_n, \ldots$ is given.

Number sequences are denoted briefly as $\{x_n\}$ $n\in N$; x_n, $n=1,2,\ldots$. Numerical sequences may be given by different methods. If the sequence $\{x_n\}$ is given by the formula that expresses its every x_n term by the n number of this term, by the formula $x_n = f(n), n\in N$ this formula is called the common term formula of the sequence $\{x_n\}$. For example, for the sequence 2, 4, 8, 16, 32, ..., $x_n = 2^n$, $n = 1, 2, 3, \ldots$; for the sequence 1, −1, 1, −1, 1,... $x_n=(-1)^{n-1}$, $n=1, 2, 3, \ldots$ is the common term formula.

It is convenient to represent some number sequences by recurrent formulas. Such formulas beginning from the second and succeeding terms express the subsequent terms by the previous terms. For example, if $a_1, a_2, \ldots, a_n, \ldots$ is a number sequence formed by the terms of number series with the first term a_1 and difference d, its any term beginning with the second one may be found by the recurrent formula $a_n=a_{n-1}+d$; $n=2,3,\ldots$. Another example of number sequence given by a recurrent formula is the Fibonacci sequence determined by the formula

$$x_1=1,\ x_2=1,\ x_n=x_{n-1}+x_{n-2};\ n=3,4,\ldots$$

Sequences may be given by other methods as well. For example, if x_n $\sqrt{2}=1{,}414213\ldots$ represents the n-th decimal sign, then,

$$x_1=4, x_2=1, x_3=4, x_4=2, x_5=1, x_6=3,\ldots$$

should be taken.

Some of the elements of the sequence, even all of them may be equal. A sequence with the equal elements is called a stationary sequence. For example 1, 1, 1,... is a stationary sequence.

As sequences are the functions with natural arguments, the notions of boundedness and monotonicity given for functions are given for the sequences in the same way. If for the sequence $\{x_n\}$ there exists a number $M(m)$such that

$$x_n \leq M\left(x_n \geq m\right), n = 1, 2, \ldots$$

Then, the sequence $\{x_n\}$ is said to be a sequence bounded from above below.

A sequence bounded from both above and below is called a bounded sequence. When $\{x_n\}$ is a bounded sequence, we can find a number $A > 0$ such that for all the members of this sequence the relation $|x_n| \leq A$ be valid.

The sequence $\{x_n\}$ is called an unbounded sequence if for any number $A > 0$ this sequence has at least one element satisfying the condition $|x_n| > A$.

For example, the sequence ...,–1, –4, –9, ..., $-n^2$ is bounded from above, but not from lower. The sequence $1, \frac{1}{2}, \frac{1}{3}, \ldots, \frac{1}{n}, \ldots$ is bounded, the sequence 1, 2, 1, 3, ..., 1, n, 1, $n + 1$,... is unbounded.

If for all the terms of the sequence $\{x_n\}$ and for terms following a certain number $x_n \leq x_{n+1} \left(x_n < x_{n+1}, x_n \geq x_{n+1}, x_n > x_{n+1}\right)$, then the sequence $\{x_n\}$ is called a non-decreasing (increasing, non-increasing, and decreasing) sequence. The non-decreasing and not increasing sequences are called monotone, the increasing and decreasing sequences are called strongly monotone sequences.

Definition. If for any number $\varepsilon > 0$ we can represent a natural number N such that for $n \geq N$,

$$|x_n - a| < \varepsilon \tag{2.4}$$

then, the number a is said to be a limit of the sequence $\{x_n\}$ and is written as $\lim\limits_{n\to\infty} x_n = a$. Condition (1) and inequality $a - \varepsilon < x_n < a + \varepsilon$ are equipotential. The interval $(a - \varepsilon, a + \varepsilon)$ is called ε -vicinity of the point a. Thus, if the number a is the limit of the sequence $\{x_n\}$, then taking a small number $\varepsilon > 0$, all terms of the sequence $\{x_n\}$ following a certain number are arranged in the ε-vicinity of the point a. For example, if for the sequence

$$1+1,\ 1+\frac{1}{2},\ 1+\frac{1}{3},\ 1+\frac{1}{4},\ \ldots,\ 1+\frac{1}{n},\ \ldots$$

with general term formula $x_n = 1 + \frac{1}{n}$ we take $\varepsilon = 0{,}1$, then for $n > 11$, taking $\varepsilon = 0{,}01$ for $n > 101$, the condition $|x_n - 1| < \varepsilon$ is satisfied.

A sequence with a limit is called a convergent sequence, without a limit, a decomposing sequence.

For $\lim_{n\to\infty} \alpha_n = 0$, $\{\alpha_n\}$ is called an infinitely decreasing sequence.

Theorem 1. The necessary and sufficient condition for the number a to be the limit of the sequence $\{x_n\}$, is representation of the terms of this sequence in the form

$$x_n = a + \alpha_n, \left(\lim_{n\to\infty} \alpha_n = 0 \right).$$

When taking any number $A > 0$ if it is possible to find a natural number N such that the condition $|x_n| > A$ be satisfied for all the elements satisfying the condition $n > N$, then $\{x_n\}$ is called on infinitely increasing sequence. Formally, it is written as $\lim_{n\to\infty} x_n = \infty$. If all the terms of the sequence $\{x_n\}$ following a certain number are positive (negative), then the following notation is used

$$\lim_{n\to\infty} x_n = +\infty \left(\lim_{n\to\infty} x_n = -\infty \right).$$

Infinitely increasing sequences are considered decomposing sequences.

Every infinitely increasing sequence is unbounded. But an unbounded sequence may be also infinitely increasing. For example, the above sequence 1, 2, 1, 3, …, 1, n, 1, $n + 1$,… is unbounded, but is not infinitely increasing.

From the set of natural numbers we take certain numbers $k_1, k_2, \ldots, k_n, \ldots$ by increasing order. The sequence

$$x_{k_1}, x_{k_2}, \ldots, x_{k_n}, \ldots$$

made of elements of the sequence $\{x_n\}$ with $k_1, k_2, \ldots, k_n, \ldots$ index elements is said to be a subsequence of the sequence $\{x_n\}$. For example, subsequence made of even number terms of the sequence 1, −1, 1, −1, 1, … is a stationary sequence −1, −1, −1,….

Every subsequence of the convergent sequence is also convergent. But the subsequence may be not convergent sequence as well. For example, the subsequence $-1, -1, -1,\ldots$ of the sequence $1, -1, 1, -1,\ldots$ is convergent, but itself is decomposing. Each subsequence of infinitely increasing sequence is infinitely increasing.

Definition. If in ε-vicinity of the point a the sequence $\{x_n\}$ has infinitely many elements, then a is said to be limit point of the sequence $\{x_n\}$.

The sequence may have only one limit, but the number of limit points may be two or more.

When a is the limit point of the sequence $\{x_n\}$, from this sequence we can separate a sequence that converges to a. The convergent sequence has only one limit point coinciding with its limit.

Every bounded sequence has at least one limit point.

The greatest (least) of the limit points is called the upper (lower) limit of the sequence and is denoted as $\overline{\lim\limits_{n\to\infty}} x_n \left(\underline{\lim\limits_{n\to\infty}} x_n \right)$.

Definition. If for any $\varepsilon > 0$ one can find a natural number N such that at $n > N$ and for arbitrary natural number p,

$$\left|x_{n+p} - x_n\right| < \varepsilon, \tag{2.5}$$

then, $\{x_n\}$ is called a fundamental sequence.

Sometimes we denote $n + p = m$ and write (2.5) in the form

$$\left|x_m - x_n\right| < \varepsilon, \quad m > N,\, n > N. \tag{2.6}$$

Theorem 2. The necessary and sufficient condition for a number series to be convergent is its fundamental property.

Every convergent sequence is bounded. Sum, difference of infinitely decreasing sequences, production of infinitely decreasing sequence and unbounded sequence is also infinitely decreasing.

When $\{x_n\}$ and $\{y_n\}$ are convergent, the sequences $\{x_n \pm y_n\}$, $\{x_n \cdot y_n\}$, $\left\{\frac{x_n}{y_n}\right\}$, and $(y_n \neq 0)$ are also convergent and the relation

$$\lim_{n\to\infty}(x_n \pm y_n) = \lim_{n\to\infty} x_n \pm \lim_{n\to\infty} y_n,$$

$$\lim_{n\to\infty}(x_n \cdot y_n) = \lim_{n\to\infty} x_n \cdot \lim_{n\to\infty} y_n,$$

$$\lim_{n\to\infty}\frac{x_n}{y_n} = \frac{\lim\limits_{n\to\infty} x_n}{\lim\limits_{n\to\infty} y_n}, \left(\lim_{n\to\infty} y_n \neq 0\right)$$

is true.

If the relations $x_n \le y_n$, $n = 1, 2, \ldots$ between the convergent sequences $\{x_n\}, \{y_n\}$ are true, then $\lim_{n\to\infty} x_n \le \lim_{n\to\infty} y_n$.

If the sequences $\{x_n\}$ and $\{y_n\}$ are convergent, have common limit a, then $\{z_n\}$ is a sequence such that for all terms following the certain number, the relation $x_n \le z_n \le y_n$ is true, then the sequence $\{z_n\}$ is convergent and $\lim_{n\to\infty} z_n = a$.

Theorem 3. A monotone bounded sequence is convergent. It is easy to verify that the sequence $\{x_n\}$ with common term formula $x_n = \left(1+\frac{1}{n}\right)^n$ is increasing $(x_{n+1} > x_n)$ and bounded $(2 \le x_n < 3)$. The limit of this sequence is called the number e

$$e = \lim_{n\to\infty}\left(1+\frac{1}{n}\right)^n.$$

e is an irrational number: $e = 2,718281828459\ldots$.

A logarithm whose base is the number e is called a natural logarithm. The natural logarithm of the number a is denoted by ln a.

Problems to be solved in auditorium

Problem 266. Write first four terms of the sequences given with common term formula:

$$1)\, x_n = \left(1+\frac{1}{n}\right)^n;\ 2)\, x_n = \frac{\sin(n\pi/2)}{n}.$$

Answer: $1)\, 2, 2\frac{1}{4}, 2\frac{10}{27}, 2\frac{113}{256}, \ldots;\ 2)\, 1, 0, -\frac{1}{3}, 0, \ldots$.

Problem 267. Which of the numbers a = 1215 and b = 12555 is the term of the sequence with common term formula $x_n = 5\cdot 3^{2n-3}$?

Answer: a.

Problem 268. Show the common term formula for the following sequences with the given initial terms:

$$1)\ 8, 14, 20, 26, 32, \ldots;\ 2)\, 2, \frac{3}{2}, \frac{4}{3}, \frac{5}{4}, \frac{6}{5}, \ldots;$$

$$3)\ 5, 7, 11, 19, 35, \ldots;\ \ 4)\ 0,3;\ 0,33;\ 0,333;\ 0,3333;\ \ldots.$$

Answer: $1)\, x_n = 2+6n; 2)\, x_n = \frac{n+1}{n};$

$3)\, x_n = 3+2^n; 4)\, x_n = \frac{1-10^{-n}}{3}.$

Problem 269. Determine which of the sequences with the given the common term formula is unbounded:

$$1)\, x_n = \frac{5n^2}{n^2+3}; 2)\, x_n = (-1)^n \frac{2n}{n+1}\sin n;$$

$$3)\, x_n = n\cos \pi n; 4)\, x_n = \frac{(-1)^n \cdot n + 10}{\sqrt{n^2+1}};$$

$$5)\, x_n = n^{\cos \pi n}.$$

Solution of 4): By the property of the summation modulus

$$\left|(-1)^n \cdot n + 10\right| \le \left|(-1)^n \cdot n\right| + 10 = n + 10.$$

On the other hand, as $\sqrt{n^2+1} > n$, we can write

$$|x_n| = \frac{\left|(-1)^n \cdot n + 10\right|}{\sqrt{n^2+1}} \le \frac{n+10}{n} = 1 + \frac{10}{n} \le 1 + 10 = 11.$$

Thus, as in any natural values of n, $|x_n| \le 11$, the given sequence is bounded.

Solution of 5): It is easy to show that a sequence with the common term formula $x_n = n^{\cos \pi n}$ is unbounded. Indeed, for a number $A > 0$ we can find an even positive number $n_0 = 2k$ such that, $2k > A$. Then we get

$$x_{n_0} = n_0^{\cos \pi n_0} = (2k)^{\cos 2k\pi} = 2k > A.$$

So, for any number $A > 0$ the given sequence has a term x_{n_0} such that $x_{n_0} > A$. This shows that a sequence with the common term formula $x_n = n^{\cos \pi n}$ is unbounded.

Answer: 1) bounded: $0 < x_n < 5$; 2) bounded: $|x_n| < 2$; 3) unbounded.

Problem 270. Prove that starting with certain number the following sequences are monotone:

$$1)\left\{\frac{n+1}{2n-1}\right\};2)\left\{\frac{100n}{n^2+16}\right\};3)\left\{\sqrt{n+2}-\sqrt{n+1}\right\};$$
$$4)\left\{3^n-2^n\right\};5)\left\{\frac{5^n}{n!}\right\}.$$

Solution of 5): Write the common term formula of the sequence:

$$x_n=\frac{5^n}{n!}\Rightarrow x_{n+1}=\frac{5^{n+1}}{(n+1)!}.$$

Let us consider the following ratio:

$$\frac{x_{n+1}}{x_n}=\frac{5^{n+1}\cdot n!}{5^n\cdot(n+1)!}=\frac{5}{n+1}.$$

For $n \geq 5$, as $\frac{5}{n+1}\leq\frac{5}{6}<1$, we get that starting with the fifth term $\frac{x_{n+1}}{x_n}<1\Rightarrow x_{n+1}<x_n$. This means that for $n \geq 5$ the given sequence is decreasing.

Answer: 1) is decreasing; 2) for $n \geq 4$ is decreasing; 3) is decreasing; 4) is increasing.

Guideline: In 1), 2), 4) look at the differences $x_n - x_{n+1}$ or $x_{n+1} - x_n$, in 3) at first show the common term in the form $x_n=\frac{1}{\sqrt{n+2}-\sqrt{n+1}}$.

Problem 271. For $x_n=\frac{1}{n+1},(n=1,2,\ldots)$, by the definition of the limit show that $\lim_{n\to\infty} x_n=0$.

Problem 272. Find the limit a of the sequence $\{x_n\}$ whose common term is determined as $x_n = 0,\underbrace{33...3}_{n}$ and take $\varepsilon = 0.001$ and find a natural number N such that for $n > N$ $|x_n-a|<\varepsilon$.

Solution. The common term formula of the sequence whose terms are determined as $x_n=0,\underbrace{33\ldots3}_{n}$, may be written in the form $x_n=\frac{1-10^{-n}}{3}$ (see problem 268, 4). Therefore we can write

$$\lim_{n\to\infty} x_n = \lim_{n\to\infty} \frac{1-10^{-n}}{3} = \frac{1-0}{3} = \frac{1}{3}.$$

Now let us try to find a number N such that $n > N$ for $\left|x_n - \frac{1}{3}\right| < 0.001$.

Taking the expression of x_n into account, we can write

$$\left|x_n - \frac{1}{3}\right| = \left|\frac{1-10^{-n}}{3} - \frac{1}{3}\right| = \frac{10^{-n}}{3}.$$

It is clear that for $n > 3$, $10^{-n} < 0.003$. Therefore, for $n > 3$, the condition $\left|x_n - \frac{1}{3}\right| < 0.001$ will be satisfied. Thus, $N = 3$ should be taken.

Problem 273. Find the limits:

$$1) \lim_{n\to\infty} \frac{5n+1}{7-9n}; 2) \lim_{n\to\infty} \frac{(n+1)^2}{2n^3};$$

$$3) \lim_{n\to\infty} \frac{\sqrt[3]{n^4+3n+1}}{n^2-1}; 4) \lim_{n\to\infty} \frac{2^n+3^n}{2^n-3^n};$$

$$5) \lim_{n\to\infty} \left(\sqrt{n^2-1}-n-1\right); 6) \lim_{n\to\infty} \frac{\sqrt{n^2+1}+\sqrt{n}}{\sqrt[3]{n^3+n}+n}.$$

Solution of 4): $\lim_{n\to\infty} \frac{2^n+3^n}{2^n-3^n} = \lim_{n\to\infty} \frac{\left(\frac{2}{3}\right)^n+1}{\left(\frac{2}{3}\right)^n-1} = \frac{0+1}{0-1} = -1;$

Solution of 5):

$$\lim_{n\to\infty} \left(\sqrt{n^2-1}-n-1\right) = \lim_{n\to\infty} \frac{\left(\sqrt{n^2-1}\right)^2-(n+1)^2}{\sqrt{n^2-1}+n+1} = \lim_{n\to\infty} \frac{n^2-1-n^2-2n-1}{\sqrt{n^2-1}+n+1} =$$

$$= \lim_{n\to\infty} \frac{-2n-2}{\sqrt{n^2-1}+n+1} = \lim_{n\to\infty} \frac{-2-\frac{2}{n}}{\sqrt{1-\frac{1}{n^2}}+1+\frac{1}{n}} = \frac{-2-0}{\sqrt{1-0}+1+0} = -1.$$

Answer: $1)-\frac{5}{9};2)0;3)0;6)\frac{1}{2}$.

Problem 274. Prove that:

$$1)\lim_{n\to\infty}\sqrt[3n]{8};2)\lim_{n\to\infty}\sqrt[n^2]{6};3)\lim_{n\to\infty}\sqrt[n]{\frac{5n+1}{n+5}};$$

$$4)\lim_{n\to\infty}\sqrt[n]{3^n+2^n}.$$

Solution of 4): $\lim_{n\to\infty}\sqrt[n]{3^n+2^n}=\lim_{n\to\infty}\sqrt[n]{3^n\left(1+\left(\frac{2}{3}\right)^n\right)}=$

$$=\lim_{n\to\infty}3\sqrt[n]{1+\left(\frac{2}{3}\right)^n}=3\lim_{n\to\infty}\left(1+\left(\frac{2}{3}\right)^n\right)^{\frac{1}{n}}=3\cdot(1+0)^0=3.$$

Answer: 1) 1; 2) 1; 3) 1.

Problem 275. Prove that $\lim_{n\to\infty}\sqrt[n]{n}=1$.

Solution. Denote $\sqrt[n]{n}-1=\alpha_n$ hence we get $\sqrt[n]{n}=1+\alpha_n$. If we can prove that $\lim_{n\to\infty}\alpha_n=0$, then by theorem 1 we can state that $\lim_{n\to\infty}\sqrt[n]{n}=1$.

It is clear that for any natural values of n, $\alpha_n\geq 0$. We raise the both hand sides of the equality $\sqrt[n]{n}=1+\alpha_n$ to power to the n-th degree and take $\alpha_n\geq 0$ into account:

$$n=(1+\alpha_n)^n=1+n\alpha_n+\frac{n(n-1)}{2!}\alpha_n^2+\ldots+\alpha_n^n>1+\frac{n(n-1)}{2}\alpha_n^2\Rightarrow$$

$$\Rightarrow n>1+\frac{n(n-1)}{2}\alpha_n^2\Rightarrow n-1>\frac{n(n-1)}{2}\alpha_n^2\Rightarrow 1>\frac{n}{2}\alpha_n^2\Rightarrow$$

$$\Rightarrow\alpha_n^2<\frac{2}{n}\Rightarrow 0\leq\alpha_n<\sqrt{\frac{2}{n}}.$$

As $\lim_{n\to\infty}\sqrt{\frac{2}{n}}=0$, from the last relation it follows that $\lim_{n\to\infty}\alpha_n=0$. Hence and from the ratio $\sqrt[n]{n}=1+\alpha$ by theorem 1 we get that, $\lim_{n\to\infty}\sqrt[n]{n}=1$.

Problem 276. Using the relation $\lim\limits_{n\to\infty}\sqrt[n]{n}=1$, find the following limits:

$$1)\ \lim_{n\to\infty}\sqrt[n]{n^2}\,;2)\ \lim_{n\to\infty}\sqrt[n^2]{n}\,;3)\ \lim_{n\to\infty}\sqrt[n]{n+3}\,;$$

$$4)\ \lim_{n\to\infty}\frac{\sqrt[n]{n^3}+\sqrt[n]{7}}{\sqrt[n]{n^2}+\sqrt[n]{3n}}.$$

Solution of 3):

$$\lim_{n\to\infty}\sqrt[n]{n+3}=\lim_{n\to\infty}(n+3)^{\frac{1}{n+3}\cdot\frac{n+3}{n}}=$$

$$=\lim_{n\to\infty}\left((n+3)^{\frac{1}{n+3}}\right)^{1+\frac{3}{n}}=\lim_{n\to\infty}\left(\sqrt[n+3]{n+3}\right)^{1+\frac{3}{n}}=1^{1+0}=1.$$

Answer: 1) 1; 2) 1; 4) 1.

Problem 277. Prove that $\lim\limits_{n\to\infty}\dfrac{n}{a^n}=0$.for $a>1$.

Solution. Since $a>1\Rightarrow a-1>0$, as in the solution of problem 275 we can write

$$a^n=(1+a-1)^n\geq\frac{n(n-1)}{2}(a-1)^2.$$

For $n \geq 2$, using $2n-n\geq 2\Rightarrow 2n-2\geq n\Rightarrow n-1\geq\frac{n}{2}$, from the above inequality we get:

$$a^n\geq\frac{n(n-1)}{2}(a-1)^2\geq\frac{n^2}{4}(a-1)^2\Rightarrow a^n\geq\frac{n^2}{4}(a-1)^2\Rightarrow$$

$$\Rightarrow\frac{n^2}{4}(a-1)^2\leq a^n\Rightarrow 0\leq\frac{n}{a^n}\leq\frac{4}{n(a-1)^2}.$$

As $\lim\limits_{n\to\infty}\dfrac{4}{n(a-1)^2}=0$, from the last relation we get $\lim\limits_{n\to\infty}\dfrac{n}{a^n}=0$, $(a>1)$.

Problem 278*. Prove that $\lim\limits_{n\to\infty}\dfrac{\log_a n}{n}=0$ for $a>1$.

Problem 279. Calculate the following limits:

$$1)\ \lim_{n\to\infty}\frac{n\lg n}{n^2-1};\ 2)\ \lim_{n\to\infty}\frac{\log_2(n+3)}{n-1{,}3}.$$

Solution of 1): Divide the denominator and numerator of the fraction under the limit sign into n^2, use $\lim\limits_{n\to\infty}\frac{\lg n}{n}=0$:

$$\lim_{n\to\infty}\frac{n\lg n}{n^2-1}=\lim_{n\to\infty}\frac{\frac{\lg n}{n}}{1-\frac{1}{n^2}}=\frac{0}{1-0}=0.$$

Answer: 2) 0.

Problem 280. Show that the sequence with common term formula $x_n=3^{\sqrt[3]{n}}$ is infinitely increasing.

Solution. In order to show that the given sequence is infinitely increasing, it is necessary to prove that for any number $M>0$ one can find a natural number N such that for all the terms of this sequence satisfying the condition $n>N$ the relation

$$x_n>M\Rightarrow 3^{\sqrt[3]{n}}>M$$

be valid. From the last inequality we get,

$$\sqrt[3]{n}>\log_3 M\Rightarrow n>(\log_3 M)^3.$$

So, if we take $N=[\log_3 M]^3$, then for $n>N$ $|x_n|>M$ (here the sign $[x]$ indicates the entire part of x). This means that the given sequence is infinitely increasing.

Problem 281. Prove that the sequence $x_n=\frac{1}{n}$, $n=1,2,...$ is fundamental.

Problem 282. Prove that the sequence with the common term formula $x_n=2+\frac{1}{2!}+\frac{1}{3!}+...+\frac{1}{n!}$ is convergent.

Guideline: Show that the sequence is monotone and bounded use theorem 3.

Problem 283. Show that the sequence given by the recurrent formula $x_1=\sqrt{2}$, $x_n=\sqrt{2+x_{n-1}}$, $n=2,3,...$ is convergent and find its limit.

Solution. It is clear that this sequence is increasing. Prove that this sequence is bounded from above. Indeed,

$$x_1=\sqrt{2}<2,\ x_2=\sqrt{2+x_1}<\sqrt{2+2}=2,$$

$x_3 = \sqrt{2+x_2} < \sqrt{2+2} = 2$, etc. Thus, for any number n, $x_n < 2$. By theorem 3 this sequence is convergent. Denote its limit by a: $\lim\limits_{n\to\infty} x_n = a$.

Raise the both hand sides of the relation $x_n = \sqrt{2+x_{n-1}}$ in square and pass to limit:

$$\lim_{n\to\infty} x_n^2 = 2 + \lim_{n\to\infty} x_{n-1} \Rightarrow a^2 = 2+a \Rightarrow a^2 - a - 2 = 0.$$

Having solved the last quadratic equation, we find $a = 2$ and $a = -1$. As $x_n > 0$, $n = 1,2,\ldots$the limit of $\{x_n\}$may not be negative. So, $\lim\limits_{n\to\infty} x_n = 2$.

Problem 284. Find the limits:

$$1)\lim_{n\to\infty}\left(1+\frac{1}{n+k}\right)^n;\ 2)\lim_{n\to\infty}\left(\frac{n}{n+1}\right)^n;$$

$$3)\lim_{n\to\infty}\left(1+\frac{1}{n}\right)^{2n};\ 4)\lim_{n\to\infty}\left(1+\frac{1}{2n}\right)^n;$$

Solution of 1): Using $\lim\limits_{n\to\infty}\left(1+\frac{1}{n}\right)^n = e$,

$$\lim_{n\to\infty}\left(1+\frac{1}{n+k}\right)^n = \lim_{n\to\infty}\left(1+\frac{1}{n+k}\right)^{n+k-k} = \lim_{n\to\infty}\left(1+\frac{1}{n+k}\right)^{n+k} \times$$

$$\times\left(1+\frac{1}{n+k}\right)^{-k} = \lim_{n+k\to\infty}\left(1+\frac{1}{n+k}\right)^{n+k} \cdot \lim_{n\to\infty}\left(1+\frac{1}{n+k}\right)^{-k} =$$

$$= e\cdot(1+0)^{-k} = e.$$

Answer: $2)e^{-1}$; $3)e^2$; $4)\sqrt{e}$.

Home tasks

Problem 285. Write the first four terms of the sequences with the common term formula:

$$1)x_n = \sin\frac{n\pi}{3};2)x_n = 2^{-n}\cos n\pi;3)x_n = \frac{\sin(n\pi/2)}{n}.$$

Answer: $1)\frac{\sqrt{3}}{2},\frac{\sqrt{3}}{2},0,-\frac{\sqrt{3}}{2},...;2)-\frac{1}{2},\frac{1}{4},-\frac{1}{8},\frac{1}{16},...;$

$3)1,0,-\frac{1}{3},0,...$.

Problem 286. Represent the common term formulas for the following sequences with the given initial terms:

$$1)-0,5;1,5;-4,5;13,9;-40,5;...;$$
$$2)1,3,1,3,1,...;\ 3)1,2,6,24,120,....$$

Answer: $1)x_n=\frac{(-3)^n}{6};2)x_n=2+(-1)^n;3)x_n=n!$.

Problem 287. Show that the sequences are bounded:

$$1)x_n=\frac{2n^2-1}{2+n^2};2)x_n=\frac{1-n}{\sqrt{n^2+1}};3)x_n=\frac{n+(-1)^n}{3n-1}.$$

Problem 288. Show that the sequences are unbounded:

$$1)x_n=(-1)^n n;2)x_n=n^2-n;3)x_n=\frac{1-n}{\sqrt{n}}.$$

Problem 289. Prove that beginning with certain number the following sequences are monotone:

$$1)x_n=\frac{3n+4}{n+2};2)x_n=n^3-6n^2;3)x_n=\lg(n+1)-\lg n.$$

Answer: 1) is increasing; 2) for $n\geq 4$ is increasing; 3) is decreasing.

Guideline: Write the common term in the form $x_n=\lg\left(1+\frac{1}{n}\right)$.

Problem 290. Using the definition of the limit of the sequence, show that $\lim\limits_{n\to\infty} {2n-1}/{2n+1}=1$.

Problem 291. Prove that the sequences with the below given common term formulas are decomposing:

$$1)x_n=(-1)^n;2)x_n=\sin\frac{n\pi}{2}.$$

Problem 292. Find the limits:

$$1)\lim_{n\to\infty}\frac{n-1}{3n};\ 2)\lim_{n\to\infty}\frac{3n^2-7n+1}{2-5n-6n^2};\ 3)\lim_{n\to\infty}\frac{(n+2)^3-(n-2)^3}{95n^3+39n};$$

$$4)\lim_{n\to\infty}\left(\sqrt{n+2}-\sqrt{n}\right);\ 5)\lim_{n\to\infty}\left(\sqrt{n^2+n}-n\right);$$

$$6)\lim_{n\to\infty}\frac{2^{n+2}+3^{n+3}}{2^n+3^n};\ 7)\lim_{n\to\infty}n^{3/2}\left(\sqrt{n^3+1}-\sqrt{n^3-2}\right).$$

Answer: $1)\frac{1}{3};2)-\frac{1}{2};3)0;4)0;5)\frac{1}{2};6)27;7)1\frac{1}{2}.$

Problem 293. Find the limits:

$$1)\ \sqrt[2n]{0.5};\ 2)\ \sqrt[n]{\frac{2n+5}{n-0.5}};\ 3)\ \sqrt[n]{5n};\ 4)\ \sqrt[n]{3n-2}.$$

Answer: 1) 1; 2) 1; 3) 1; 4) 1.

Problem 294. Show that the sequence given by the recurrent formula $x_1=13,\ x_{n+1}=\sqrt{12+x_n}$ is convergent and find its limit.

Answer: 4.

Problem 295. Find the limits:

$$1)\lim_{n\to\infty}\left(\frac{2^n+1}{2^n}\right)^{2^n};2)\lim_{n\to\infty}\left(1-\frac{1}{n}\right)^n;$$

$$3)\lim_{n\to\infty}\left(1+\frac{1}{n(n+2)}\right)^n;$$

$$4)\lim_{n\to\infty}\left(\frac{2^n+3}{2^n+1}\right)^n;5)\lim_{n\to\infty}\left(\frac{n^2+1}{n^2-2}\right)^n;$$

$$6)\lim_{n\to\infty}\left(\frac{n^2+n}{n^2+2n+2}\right)^n;\ 7)\lim_{n\to\infty}\left(\frac{n^2-n+1}{n^2+n+1}\right)^n.$$

Answer: 1) e; 2) e^{-1}; 3) 1; 4) 1; 5) e^3; 6) e^{-1}; 7) e^{-2}.

2.3 LIMIT OF FUNCTION. INFINITELY DECREASING FUNCTIONS

If in any neighborhood of the point a the set X has an element different from a, then a is said to be a limit point of the set X. The limit point of the set may either enter into this set or may not.

Assume that the function $y = f(x)$ was determined in the set X, a is the limit point of the set X. As a is the limit point of the set X, from the set X, we can separate the sequence

$$x_1, x_2, \ldots, x_n, \ldots \tag{2.7}$$

that converges to a and its elements are different from a. It is possible to separate from the set X many sequences in the form of (2.7) and convergent to a. If all the sequences of values

$$f(x_1),\ f(x_2), \ldots, f(x_n) \ldots \tag{2.8}$$

of the function $f(x_1)$ corresponding to all sequences in the form of (2.7) converge to the same number b, then the number b is called the limit of the function $y = f(x)$ at the point a.

The fact that the number b is the limit of the function $y = f(x)$ at the point a is denoted as

$$\lim_{x \to a} f(x) = b .$$

It is clear that the function that has a limit at the point a not to be determined at this point.

The above definition is called the definition of limit "in the terms of sequence" or in "Heine's sense."

The limit of a function may also have the following definition.

Definition. If for the numbers $a \times b$ and arbitrary $\varepsilon > 0$ one can find a number $\delta > 0$ such that the relation $|f(x) - b| < \varepsilon$ be satisfied at all the values of x from X and satisfying the condition $0 < |x - a| < \delta$, then the number b is called the limit of the function $y = f(x)$ at the point a.

This definition of the limit is called the definition "in the terms of $\varepsilon - \delta$" or in "Cauchy sense."

The definitions of the limit in Cauchy and Heine's sense are equipotential.

There are such functions $f(x)$ that their x argument takes values from the set X and by approximating to a the values of the function $f(x)$ increase unboundedly in absolute value. In this case, formally it is said that at the point $x = a$ the limit of the function equals infinity and this is written as $\lim_{x \to a} f(x) = \infty$. The function, whose limit equals infinity, is called an infinitely increasing function at this point. The fact that the limit of the function is infinity may be more exactly expressed as follows.

Definition. It for any number $M > 0$ there is a number $\delta > 0$ that in all the values of x from X satisfying the inequality $0 < |x - a| < \delta$, $|f(x)| > M$, then it is said that the limit of $f(x)$ at the point a equals infinity.

When the values of the function whose limit equals infinity at the point a, are positive (negative) at certain neighborhood of the point a (except the point a), we write,

$$\lim_{x \to a} f(x) = +\infty \quad \left(\lim_{x \to a} f(x) = -\infty\right).$$

Definition. When the number b is given by any small number $\varepsilon > 0$, it is possible to find a number $M > 0$ that in all the values of x satisfying the condition $|x| > M$,

$$|f(x) - b| < \varepsilon,$$

then, the number b is called the limit of the function $y = f(x)$ provided $x \to \infty$. This is written as

$$\lim_{x \to \infty} f(x) = b.$$

When all the elements of sequence (1) convergent to a are small (great) than a, if all the sequences in the form (2) converge to b, then the number b is called the left (right) limit of the function $f(x)$ at the point a.

Express this definition in the terms of "$\varepsilon - \delta$."

Definition. If for the numbers a, b and any $\varepsilon > 0$ it is possible to find a number $\delta > 0$ such that at all the values of x from X satisfying the condition

$$0 < a - x < \delta \quad (0 < x - a < \delta),$$

the inequality

$$|f(x) - b| < \varepsilon,$$

is satisfied, then the number b is called the left (right) limit of the function $y = f(x)$ at the point a.

We can use any of the symbols for the right limit of $f(x)$ at the point a

$$\lim_{x\to a+0} f(x),\ \lim_{\substack{x\to a\\(x>a)}} f(x),\ f(a+0),$$

for the left limit

$$\lim_{x\to a-0} f(x),\ \lim_{\substack{x\to a\\(x<a)}} f(x),\ f(a-0).$$

For the function to have a limit at the point, the existence and equality of both of left and right limits at this point is the necessary and sufficient condition.

A function cannot have two different limits at one point.

The function that has a zero limit at the point a is called an infinitely decreasing function at this point.

The necessary and sufficient condition for the number b to be a limit of the function $f(x)$ at the point $x = a$ is that the difference $\alpha(x) = f(x) - b$ should be infinitely decreasing at the point $x = a$.

When the function $f(x)$ is infinitely decreasing at the point a, then $1/f(x)$ is infinitely increasing, and vice versa.

The product of an infinitely decreasing function and a bounded function is an infinitely decreasing function.

The sum of finitely many decreasing functions is an infinitely decreasing function.

If the functions $f(x)$ and $g(x)$ have limits at the same point a, then the functions,

$$f(x)\pm g(x), f(x)\cdot g(x), \frac{f(x)}{g(x)},\ \left(\lim_{x\to a} g(x)\neq 0\right),$$

also have limits at the point a and the following equalities are valid

$$\lim_{x\to a}\left(f(x)\pm g(x)\right) = \lim_{x\to a} f(x) \pm \lim_{x\to a} g(x),$$

$$\lim_{x\to a}\left(f(x)\cdot g(x)\right) = \lim_{x\to a}(x)\cdot \lim_{x\to a} g(x),$$

$$\lim_{x\to a}\frac{f(x)}{g(x)} = \frac{\lim_{x\to a} f(x)}{\lim_{x\to a} g(x)},\ \left(\lim_{x\to a} g(x)\neq 0\right).$$

If the function $f(x)$ that has a finite limit at the point a satisfies the inequality $f(x) > q$ in all values $(x \neq a)$ from certain neighborhood of a, then

$$\lim_{x \to a} f(x) \geq q .$$

Assume that at the same point a we are given two infinitely decreasing functions $\alpha(x)$ and $\beta(x)$:

$$\lim_{x \to a} \alpha(x) = 0, \ \lim_{x \to a} \beta(x) = 0.$$

If $\lim\limits_{x \to a} \dfrac{\alpha(x)}{\beta(x)} = 0$, the function $\alpha(x)$ is called a highest order infinitely decreasing function with respect to the function $\beta(x)$ and is written in the form $\alpha(x) = o(\beta(x))$. The last notation is read as follows: "$\alpha(x)$ equals small $\beta(x)$."

If there exists $\lim\limits_{x \to a} \dfrac{\alpha(x)}{\beta(x)}$ and is nonzero, then $\alpha(x)$ and $\beta(x)$ are called infinitely decreasing functions of the same order.

For $\lim\limits_{x \to a} \dfrac{\alpha(x)}{(\beta(x))^m} = c \neq 0$, the infinitely decreasing $\alpha(x)$ is called an infinitely decreasing function of m-th order with respect to the infinitely decreasing function $\beta(x)$. In this definition, instead of $\beta(x)$ having taken the infinitely decreasing $(x - a)$, for

$$\lim_{x \to a} \frac{\alpha(x)}{(x-a)^m} = c \neq 0 ,$$

$\alpha(x)$ is called an infinitely decreasing function of m-th order at the point a.

For $\lim\limits_{x \to a} \dfrac{\alpha(x)}{\beta(x)} = 1$, the $\alpha(x)$ and $\beta(x)$ are called equivalent infinitely decreasing functions.

If provided $x \to a$, when the $\alpha(x)$ and $\beta(x)$ are equivalent infinitely decreasing functions, we use the notation $\alpha(x) \sim \beta(x)$ $x \to a$.

If provided $x \to a$, $\gamma(x) \sim \alpha(x)$, $\delta(x) \sim \beta(x)$, then the equality

$$\lim_{x \to a} \frac{\alpha(x)}{\beta(x)} = \lim_{x \to a} \frac{\gamma(x)}{\delta(x)}$$

is valid.

When provided $x \to a$, when $\alpha(x)$ is an infinitely decreasing function, the following functions are equivalent infinitely decreasing functions:

$$1) \sin\alpha(x) \sim \alpha(x);\ 2) tg\alpha(x) \sim \alpha(x);$$

$$3) 1-\cos\alpha(x) \sim \frac{(\alpha(x))^2}{2};\ 4) \arcsin\alpha(x) \sim \alpha(x);$$

$$5) arctg\alpha(x) \sim \alpha(x);\ 6) \ln(1+\alpha(x)) \sim \alpha(x);$$

$$7) a^{\alpha(x)} - 1 \sim \alpha(x)\ln a, (a > 0);\ 8) e^{\alpha(x)} - 1 \sim \alpha(x);$$

$$9) (1+\alpha(x))^p - 1 \sim p\alpha(x);\ 10) \sqrt[n]{1+\alpha(x)} - 1 \sim \frac{\alpha(x)}{n}.$$

For $a = e$ 8) may be considered as a special case of 7), for $p = \frac{1}{n}$, 10) may be considered as a special case of 9).

In calculation of limits the following two limits play an important part:

$$\lim_{x\to 0} \frac{\sin x}{x} = 1, \tag{2.9}$$

$$\lim_{x\to 0} (1+x)^{\frac{1}{x}} = e. \tag{2.10}$$

Sometimes it is convenient to use the formulas obtained from (2.10)

$$\lim_{x\to\infty} \left(1+\frac{1}{x}\right)^x = e, \tag{2.11}$$

$$\lim_{x\to 0} \frac{\log_a(1+x)}{x} = \frac{1}{\ln a},\ 0 < a \neq 1, \tag{2.12}$$

$$\lim_{x\to 0} \frac{a^x - 1}{x} = \ln a,\ a > 0, \tag{2.13}$$

Taking $a = e$, formulas (2.12) and (2.13) accept the form

$$\lim_{x\to 0} \frac{\ln(1+x)}{x} = 1, \tag{2.14}$$

$$\lim_{x\to 0} \frac{e^x - 1}{x} = 1, \tag{2.15}$$

respectively.

When the $\lim\limits_{x\to a} u(x) > 0$ and $\lim\limits_{x\to a} v(x)$ limits are finite, there exists the limit $\lim\limits_{x\to a} u(x)^{v(x)}$, $(u(x) > 0)$ and the formula

$$\lim_{x\to a} u(x)^{v(x)} = \left(\lim_{x\to a} u(x)\right)^{\lim\limits_{x\to a} \vartheta(x)} \tag{2.16}$$

is valid.

Problems to be solved in auditorium

Problem 296. Find the limit of the function $f(x) = \dfrac{x^2-16}{x^2-4x}$ at the point $x = 4$ by both definitions of the limit of a function.

Solution. At first, based on the Heine's definition, find the limit of the function at the point $x = 4$. For that, take a sequence $\{x_n\}$ whose all elements are from certain neighborhood of the point $x = 4$, for example, from the interval (3;5) that $x_n \neq 4$, $n = 1, 2, \ldots$, and $\lim\limits_{n\to\infty} x_n = 4$. Make the sequence $\{f(x_n)\}$ of appropriate values of the function $f(x)$ and calculate its limit:

$$\lim_{n\to\infty} f(x_n) = \lim_{n\to\infty} \frac{x_n^2-16}{x_n^2-4x_n} = \lim_{n\to\infty} \frac{x_n+4}{x_n} = \frac{\lim\limits_{n\to\infty} x_n + 4}{\lim\limits_{n\to\infty} x_n} = \frac{4+4}{4} = 2 .$$

As is seen $\lim\limits_{n\to\infty} f(x_n) = 2$ is independent on the choice of the sequence satisfying the condition $\lim\limits_{n\to\infty} f(x_n) = 4$. Therefore, according to the Heine definition of the limit, we get $\lim\limits_{x\to 4} f(x) = 2$.

Now use the Cauchy definition of the limit and show that $\lim\limits_{x\to 4} \dfrac{x^2-16}{x^2-4x} = 2$. For that let us consider the function $f(x)$ at certain neighborhood of the point $x = 4$, for example, on the interval (2; 5).

Take any number $\varepsilon > 0$ and for $x \neq 4$ transform $|f(x)-2|$ as follows:

$$\left|\frac{x^2-16}{x^2-4x} - 2\right| = \left|\frac{x+4}{x} - 2\right| = \frac{|x-4|}{|x|} = \frac{|x-4|}{x} .$$

As $x \in (2;5)$, we take $|x| = x$ and can write $\frac{1}{x} < \frac{1}{2}$. Then we get

$$\left|\frac{x^2-16}{x^2-4x}-2\right| < \frac{|x-4|}{2}.$$

From the last inequality, it is clear that if we take $\delta = 2\varepsilon$, for all the numbers $x \in (2;5)$ satisfying the inequality $0 < |x-4| < \delta$,

$$\left|\frac{x^2-16}{x^2-4x}-2\right| < \frac{\delta}{2} = \varepsilon .$$

This means that by definition of limit in the Cauchy sense, $\lim\limits_{x\to 4} f(x) = 2$.

Problem 297. Prove that the following limits are absent:

$$1) \lim_{x\to 1} \sin\frac{1}{x-1};\ 2) \lim_{x\to 0} 2^{\frac{1}{x}}.$$

Solution of 1): Let us consider two sequences $\{x_n\}$ and $\{x_n'\}$ that converge to 1 and have the following common term formulas:

$$x_n = 1 + \frac{1}{n\pi}\ ,\quad x_n' = 1 + \frac{2}{(4n+1)\pi}.$$

Find the sequence of values appropriate to this number sequences $\{x_n\}$ and $\{x_n'\}$ of the function $f(x) = \sin\frac{1}{x-1}$:

$$f(x_n) = \sin\frac{1}{1+\frac{1}{n\pi}-1} = \sin n\pi = 0,$$

$$f(x_n') = \sin\frac{1}{1+\frac{2}{(4n+1)\pi}-1} = \sin\frac{(4n+1)\pi}{2} = \sin\left(2n\pi + \frac{\pi}{2}\right) = 1,\ n = 1, 2\dots .$$

As is seen, the sequences $\{f(x_n)\}$ and $\{f(x_n')\}$ have different limits. So, the limit $\lim\limits_{x\to 1}\sin\frac{1}{x-1}$ is absent.

Guideline: For showing that 2) has no limit, take the sequences with common term formulas $x_n = \pi m$ and $x_n' = 2\pi n + \frac{\pi}{2}$.

Problem 298. Find the limits:

$$1)\lim_{x\to 1}\frac{x^2+4x-5}{x^2-1};\ 2)\lim_{x\to 0}\frac{x^7+5x^6+4x^3}{x^7+2x^3};$$

$$3)\lim_{x\to 6}\frac{\sqrt{x-2}-2}{x-6};\ 4)\lim_{x\to 1}\frac{1-\sqrt[3]{x}}{1-\sqrt[5]{x}}.$$

Solution of 4): Using the formula

$$a^n - b^n = (a-b)\left(a^{n-1}+a^{n-2}b+\ldots+ab^{n-1}+b^{n-1}\right),$$

we get

$$\lim_{x\to 1}\frac{1-\sqrt[3]{x}}{1-\sqrt[5]{x}} =$$

$$=\lim_{x\to 1}\frac{\left(1-\sqrt[3]{x}\right)\left(1+\sqrt[3]{x}+\sqrt[3]{x^2}\right)\left(1+\sqrt[5]{x}+\sqrt[5]{x^2}+\sqrt[5]{x^3}+\sqrt[5]{x^4}\right)}{\left(1-\sqrt[5]{x}\right)\left(1+\sqrt[5]{x}+\sqrt[5]{x^2}+\sqrt[5]{x^3}+\sqrt[5]{x^4}\right)\left(1+\sqrt[3]{x}+\sqrt[3]{x^2}\right)}=$$

$$=\lim_{x\to 1}\frac{(1-x)\left(1+\sqrt[5]{x}+\sqrt[5]{x^2}+\sqrt[5]{x^3}+\sqrt[5]{x^4}\right)}{(1-x)\left(1+\sqrt[3]{x}+\sqrt[3]{x^2}\right)}=$$

$$=\lim_{x\to 1}\frac{1+\sqrt[5]{x}+\sqrt[5]{x^2}+\sqrt[5]{x^3}+\sqrt[5]{x^4}}{1+\sqrt[3]{x}+\sqrt[3]{x^2}}=$$

$$=\frac{1+1+1+1+1}{1+1+1}=\frac{5}{3}.$$

Answer: 1) 3; 2) 2; 3) $\frac{1}{4}$.

Problem 299. Find the limits:

$$1)\lim_{x\to\infty}\frac{3x-1}{x^2+1};2)\lim_{x\to\infty}\frac{\sqrt{x}-6x}{3x+1};3)\lim_{x\to\infty}\left(\sqrt{x^2-1}-\sqrt{x^2+1}\right);$$

$$4)\lim_{x\to\infty}\left(\sqrt{4x^4+13x^2-7}-2x^2\right);5)\lim_{x\to\infty}\left(\sqrt{x^2+\sqrt{x^2+\sqrt{x^2}}}-\sqrt{x^2}\right).$$

Solution of 5):

$$\lim_{x\to\infty}\left(\sqrt{x^2+\sqrt{x^2+\sqrt{x^2}}}-\sqrt{x^2}\right)=\lim_{x\to\infty}\frac{x^2+\sqrt{x^2+\sqrt{x^2}}-x^2}{\sqrt{x^2+\sqrt{x^2+\sqrt{x^2}}}+\sqrt{x^2}}=$$

$$=\lim_{x\to\infty}\frac{\sqrt{x^2+\sqrt{x^2}}}{\sqrt{x^2+\sqrt{x^2+\sqrt{x^2}}}+\sqrt{x^2}}=\lim_{x\to\infty}\frac{\sqrt{1+\frac{1}{\sqrt{x^2}}}}{\sqrt{1+\frac{1}{\sqrt{x^2}}\cdot\sqrt{1+\frac{1}{\sqrt{x^2}}}}+1}=$$

$$=\frac{\sqrt{1+0}}{\sqrt{1+0\cdot\sqrt{1+0}}+1}=\frac{1}{2}.$$

Answer: 1) 0; 2) −2; 3) 0; 4) 3.25.

Problem 300. Find the limits:

$$1)\lim_{x\to1}\frac{2x-2}{\sqrt[3]{26+x}-3};2)\lim_{x\to-1}\frac{1+\sqrt[3]{x}}{1+\sqrt[5]{x}};3)\lim_{x\to\frac{\pi}{6}}\frac{\sin\left(x-\frac{\pi}{6}\right)}{\sqrt{3}-2\cos x}.$$

Solution of 1): Take $26 + x = z^3$. Then it is clear that provided $x \to 1$, $z \to 3$. As from above substitution we get $x = z^3 - 26$,

$$\lim_{x\to1}\frac{2x-2}{\sqrt[3]{26+x}-3}=\lim_{z\to3}\frac{2\left(z^3-26\right)-2}{\sqrt[3]{z^3}-3}=\lim_{z\to3}\frac{2z^3-54}{z-3}=$$

$$=\lim_{z\to3}\frac{2\left(z^3-27\right)}{z-3}=2\lim_{z\to3}\frac{\left(z-3\right)\left(z^2+3z+9\right)}{\left(z-3\right)}=2\lim_{z\to3}\left(z^2+3z+9\right)=54.$$

Answer: 2) $\frac{5}{3}$. ***Guideline:*** take $x = z^{15}$;

3) 1. ***Guideline:*** take $x-\frac{\pi}{6}=z$.

Problem 301. Using the relations $\lim_{x\to0}\frac{\sin x}{x}=1$, find the following limits:

$$1)\lim_{x\to 0}\frac{\sin 4x}{x};2)\lim_{x\to 0}\frac{\sin^2\frac{x}{2}}{x^2};3)\lim_{x\to 0}\frac{\sin 3x}{\sqrt{x+2}-\sqrt{2}};$$

$$4)\lim_{x\to 0}\frac{1-\cos 2x+tg^2x}{x\sin x};5)\lim_{x\to 0}\frac{\cos 3x^3-1}{\sin^6 2x};$$

$$6)\lim_{x\to\pi}\frac{\sin x}{\pi^2-x^2};7)\lim_{x\to 0}\frac{\sin x}{\sin 6x-\sin 7x};$$

$$8)\lim_{x\to 0}\frac{arctgx}{x};9)\lim_{x\to 0}\frac{\arcsin 2x}{x}.$$

Solution of 5):

$$\lim_{x\to 0}\frac{\cos 3x^3-1}{\sin^6 2x}=-\lim_{x\to 0}\frac{2\sin^2\frac{3x^3}{2}}{\sin^6 2x}=$$

$$=-2\lim_{x\to 0}\frac{\left(\frac{\sin\frac{3x^3}{2}}{\frac{3x^3}{2}}\right)^2\cdot\frac{9x^6}{4}}{\left(\frac{\sin 2x}{2x}\right)^6\cdot 2^6x^6}=-\frac{9}{128}\cdot\frac{\lim\limits_{y\to 0}\left(\frac{\sin y}{y}\right)^2}{\lim\limits_{z\to 0}\left(\frac{\sin z}{z}\right)^6}=-\frac{9}{128}.$$

Solution of 8): Denote $arctgx = y$. Then under the condition $x = tgy$ and $x \to 0, y \to 0$. Therefore

$$\lim_{x\to 0}\frac{arctgx}{x}=\lim_{y\to 0}\frac{y}{tgy}=\lim_{y\to 0}\frac{y}{\frac{\sin y}{\cos y}}=\lim_{y\to 0}\frac{\cos y}{\frac{\sin y}{y}}=\frac{1}{1}=1.$$

Answer: $1)4;2)\frac{1}{4};3)6\sqrt{2};4)3;6)\frac{1}{2\pi};7)-1;9)2.$

Problem 302. Using the relation $\lim\limits_{x\to\infty}\left(1+\frac{1}{x}\right)^x=e$ or $\lim\limits_{x\to 0}(1+x)^{\frac{1}{x}}=e$, find the following limits:

$$1)\lim_{x\to\infty}\left(\frac{x+1}{x-1}\right)^{x};2)\lim_{x\to\infty}\left(\frac{2x+3}{2x+1}\right)^{x+1};3)\lim_{x\to 0}(1+tgx)^{ctgx};$$

$$4)\lim_{x\to 0}(\cos x)^{\frac{1}{\sin^2 x}};5)\lim_{x\to a}\left(\frac{\sin x}{\sin a}\right)^{\frac{1}{x-a}};6)\lim_{x\to\frac{\pi}{4}}(tgx)^{tg2x}.$$

Solution of 1):

$$\lim_{x\to\infty}\left(\frac{x+1}{x-1}\right)^{x}=\lim_{x\to\infty}\left(\frac{(x-1)+2}{x-1}\right)^{x}=\lim_{x\to\infty}\left(1+\frac{2}{x-1}\right)^{x}=$$

$$=\lim_{x\to\infty}\left(1+\frac{2}{x-1}\right)^{\frac{x-1}{2}\cdot\frac{2x}{x-1}}=\lim_{x\to\infty}\left(\left(1+\frac{2}{x-1}\right)^{\frac{x-1}{2}}\right)^{\frac{2}{1-\frac{1}{x}}}=e^{\frac{2}{1-0}}=e^{2}.$$

Solution of 5):

$$\lim_{x\to a}\left(\frac{\sin x}{\sin a}\right)^{\frac{1}{x-a}}=\lim_{x\to a}\left(1+\frac{\sin x-\sin a}{\sin a}\right)^{\frac{1}{x-a}}=$$

$$=\lim_{x\to a}\left(1+\frac{\sin x-\sin a}{\sin a}\right)^{\frac{\sin a}{\sin x-\sin a}\cdot\frac{\sin x-\sin a}{(\sin a)(x-a)}}=$$

$$=\lim_{x\to a}\left(\left(1+\frac{\sin x-\sin a}{\sin a}\right)^{\frac{\sin a}{\sin x-\sin a}}\right)^{\frac{2\sin\frac{x-a}{2}\cos\frac{x+a}{2}}{(x-a)\sin a}}=$$

$$=\lim_{x\to a}\left(\left(1+\frac{\sin x-\sin a}{\sin a}\right)^{\frac{\sin a}{\sin x-\sin a}}\right)^{\frac{\sin\frac{x-a}{2}}{\frac{x-a}{2}}\cdot\frac{\cos\frac{x+a}{2}}{\sin a}}=e^{1\cdot\frac{\cos a}{\sin a}}=e^{ctga}.$$

Answer: $2)e;3)e;4)e^{-\frac{1}{2}};6)e^{-1}$.

Problem 303. Use formulas (2.12)–(2.15) and calculate the following limits:

$$1)\lim_{x\to 0}\frac{e^{-2x}-1}{x};2)\lim_{x\to e}\frac{\ln x-1}{x-e};3)\lim_{x\to\frac{\pi}{4}}\frac{\ln tgx}{\cos 2x};4)\lim_{x\to 0}\frac{\ln(1+x)}{3^x-1}.$$

Solution of 2): If we write the function under the limit sign in the form

$$\frac{\ln x-1}{x-e}=\frac{\ln x-\ln e}{e\left(\frac{x}{e}-1\right)}=\frac{\ln\frac{x}{e}}{e\left(\frac{x}{e}-1\right)},$$

and denote by $\frac{x}{e}-1=z$, provided $x\to e, z\to 0$. Taking these into account, we use (2.14) and get

$$\lim_{x\to e}\frac{\ln x-1}{x-e}=\lim_{x\to e}\frac{\ln\frac{x}{e}}{e\left(\frac{x}{e}-1\right)}=\lim_{z\to 0}\frac{\ln(1+z)}{ez}=\frac{1}{e}\cdot$$

Solution of 3): Substitute $tgx=y$. Then, as we can write, $x=arctg\, y$

$$\cos 2x=\cos 2(arctg\, y)=\frac{1-tg^2(arctg\, y)}{1+tg^2(arctg\, y)}=\frac{1-y^2}{1+y^2}$$

As provided $x\to\frac{\pi}{4}, y, y\to 1,$

$$\lim_{x\to\frac{\pi}{4}}\frac{\ln tgx}{\cos 2x}=\lim_{y\to 1}\frac{\ln y}{\frac{1-y^2}{1+y^2}}=-\lim_{y\to 1}\frac{\ln(1+y-1)}{(y-1)}\cdot\frac{1+y^2}{1+y}=$$

$$=-\lim_{z\to 0}\frac{\ln(1+z)}{z}\cdot\lim_{y\to 1}\frac{1+y^2}{1+y}=-1\cdot\frac{1+1}{1+1}=-1.$$

Answer: 1) –2. ***Guideline:*** Use (2.15);

4) $\frac{1}{\ln 3}$. ***Guideline:*** Use (2.13) and (2.14).

Problem 304. Calculate the limits:

$$1)\lim_{x\to 1}\left(\frac{1+x}{2+x}\right)^{\frac{1-\sqrt{x}}{1-x}};2)\lim_{x\to\infty}\left(\frac{2x^2+3}{2x^2+5}\right)^{8x^2+3}.$$

Solution of 2): Denote $\frac{2x^2+3}{2x^2+5}=u(x),\ 8x^2+3=v(x)$.

As

$$\lim_{x\to\infty}u(x)=\lim_{x\to\infty}\frac{2+\frac{3}{x^2}}{2+\frac{5}{x^2}}=\frac{2+0}{2+0}=1,\ \lim_{x\to\infty}v(x)=\lim_{x\to\infty}\left(8x^2+3\right)=\infty,$$

the given limit cannot be calculated by formula (2.16).

In the case $\lim_{x\to a}u(x)=1,\ \lim_{x\to a}v(x)=\infty$ for calculation of the limit $\lim_{x\to a}u(x)^{v(x)}$, we use the formula

$$\lim_{x\to a}u(x)^{v(x)}=\lim_{x\to a}\left(\left(1+\left(u(x)-1\right)\right)^{\frac{1}{u(x)-1}}\right)^{v(x)\left(u(x)-1\right)}=e^{\lim\limits_{x\to a}v(x)(u(x)-1)}. \quad (2.17)$$

For this example, we must use the notation of (2.17) in the form

$$\lim_{x\to\infty}u(x)^{v(x)}=e^{\lim\limits_{x\to\infty}v(x)(u(x)-1)}.$$

Having substituted the expressions of $u(x)$ and $v(x)$, we get

$$\lim_{x\to\infty}v(x)(u(x)-1)=\lim_{x\to\infty}\left(8x^2+3\right)\left(\frac{2x^2+3}{2x^2+5}-1\right)=$$

$$=\lim_{x\to\infty}\left(8x^2+3\right)\cdot\left(-\frac{2}{2x^2+5}\right)=-2\lim_{x\to\infty}\frac{8+\frac{3}{x^2}}{2+\frac{5}{x^2}}=-8,$$

$$\lim_{x\to\infty}u(x)^{v(x)}=\lim_{x\to\infty}\left(\frac{2x^2+3}{2x^2+5}\right)^{8x^2+3}=e^{\lim\limits_{x\to\infty}v(x)(u(x)-1)}=e^{-8}.$$

Answer: 1) $\sqrt{\frac{2}{3}}$.

Guideline: Use formula (2.16).

Problem 305. Find the represented right and left limits of the following functions:

$$1)\, f(x)=\begin{cases}1-2x, \; x<0,\\ 2^x, \; 0\le x\le 2,\\ 1, \qquad x>2,\end{cases} \lim_{x\to 0\pm 0} f(x), \lim_{x\to 2\pm 0} f(x);$$

$$2)\, f(x)=\frac{x-3}{|x-3|}, \lim_{x\to 3\pm 0} f(x); 3)\, f(x)=\frac{2+x}{4-x^2}, \lim_{x\to 2\pm 0} f(x).$$

Answer: 1) $\lim\limits_{x\to 0-0} f(x)=\lim\limits_{x\to 0+0} f(x)=1;\ \lim\limits_{x\to 2-0} f(x)=4,$

$\lim\limits_{x\to 2+0} f(x)=1,$ 2) $\lim\limits_{x\to 3-0} f(x)=-1,\ \lim\limits_{x\to 3+0} f(x)=1;$

3) $\lim\limits_{x\to 2-0} f(x)=\infty,\ \lim\limits_{x\to 2+0} f(x)=-\infty.$

Problem 306. Prove that the function $f(x)=arctg\frac{1}{x}$ has right and left limits at the point $x = 0$, but this function has no limit at this point.

Problem 307. Show that the following functions are infinitely increasing at the given points:

$$1) \text{ at the point } f(x)=\frac{2x-4}{x^2+5},\ x=2;$$

$$2) \text{ at the point } f(x)=(x-1)^2\sin^3\frac{1}{x-1},\ x=1;$$

$$3)\ f(x)=\frac{\sin x}{x}, \text{ provided } x\to\infty.$$

Solution of 2): Consider $f(x)$ as a product of the functions $\varphi(x) = (x-1)^2$ and $\psi(x)=\sin^3\frac{1}{x-1}$. As $\lim\limits_{x\to 1}\varphi(x)=\lim\limits_{x\to 1}(x-1)^2=0$, the function $\varphi(x)$ is infinitely decreasing at the point $x = 1$, as $|\psi(x)|=\left|\sin^3\frac{1}{x-1}\right|\le 1, (x\ne 1)$, $\psi(x)$

is bounded. As the product of an infinitely decreasing function and a bounded function is infinitely decreasing, the function

$$f(x) = \varphi(x) \cdot \psi(x) = (x-1)^2 \sin^3 \frac{1}{x-1}$$

is infinitely decreasing at the point $x = 1$.

Problem 308. Compare the infinitely decreasing functions at the point $x = 0$:

$$1)\, \alpha(x) = 5x^2 + 2x^5,\ \beta(x) = 3x^2 + 2x^3;$$
$$2)\, \alpha(x) = x\sin^2 x,\ \beta(x) = 2x\sin x;$$
$$3)\, \alpha(x) = x\ln(1+x),\ \beta(x) = x\sin x.$$

Solution of 3): As $\lim\limits_{x\to 0} \alpha(x) = \lim\limits_{x\to 0} x\ln(1+x) = 0$,

$\lim\limits_{x\to 0} \beta(x) = \lim\limits_{x\to 0} x\sin x = 0$, the both functions $\alpha(x)$ and $\beta(x)$ are infinitely decreasing at the point $x = 0$. In order to compare these infinitely decreasing functions, we must calculate the limits $\lim\limits_{x\to 0} \dfrac{\alpha(x)}{\beta(x)}$.

$$\lim_{x\to 0} \frac{\alpha(x)}{\beta(x)} = \lim_{x\to 0} \frac{x\ln(1+x)}{x\sin x} = \lim_{x\to 0} \frac{\dfrac{\ln(1+x)}{x}}{\dfrac{\sin x}{x}} = \frac{1}{1} = 1.$$

We use formula (2.9) and (2.14). As $\lim\limits_{x\to 0} \dfrac{\alpha(x)}{\beta(x)} = 1$, the functions $\alpha(x)$ and $\beta(x)$ are equivalent infinitely decreasing at the point $x = 0$.

Answer: 1) are infinitely decreasing functions of same order; 2) $\alpha = o(\beta)$.

Problem 309. Determine the order of the infinitely decreasing function $\alpha(x)$ with respect to the infinitely decreasing function $\beta(x)$:

$$\alpha)\, \alpha(x) = 1 - \cos x, \beta(x) = x, x \to 0 \text{ provided,}$$
$$б)\, \alpha(x) = (x-1)^3 + (x-1)^4, \beta(x) = x - 1, x \to 1 \text{ provided.}$$

Solution of a): $\lim\limits_{x\to 0} \alpha(x) = \lim\limits_{x\to 0} (1 - \cos x) = 0$, as $\lim\limits_{x\to 0} \beta(x) = \lim\limits_{x\to 0} x = 0$ the functions $\alpha(x)$ and $\beta(x)$ at infinitely decreasing at the point $x = 0$.

As $\lim\limits_{x\to 0}\dfrac{\alpha(x)}{(\beta(x))^2}=\lim\limits_{x\to 0}\dfrac{1-\cos x}{x^2}=\lim\limits_{x\to 0}\dfrac{2\sin^2\frac{x}{2}}{x^2}=\dfrac{1}{2}\lim\limits_{x\to 0}\left(\dfrac{\sin\frac{x}{2}}{\frac{x}{2}}\right)^2=\dfrac{1}{2}$,

$\alpha(x)$ is a second-order infinitely decreasing function with respect to $\beta(x)$.

Answer:2) $\alpha(x)$ is a third-order infinitely decreasing function with respect to $\beta(x)$.

Problem 310. Show that the functions $\alpha(x)=\sqrt{1+x}-1$ and $\beta(x)=\frac{1}{2}x$ are equivalent infinitely decreasing functions at the point $x = 0$. Using this equivalence find the approximate values of the numbers 1) $\sqrt{106}$; 2) $\sqrt{912}$; 3) $\sqrt{1672}$.

Solution. At first, note that as $\lim\limits_{x\to 0}\left(\sqrt{1+x}-1\right)=0$, $\lim\limits_{x\to 0}\left(\frac{1}{2}x\right)=0$, the functions $\alpha(x)$ and $\beta(x)$ are infinitely decreasing at the point $x = 0$. Show that they are equivalent infinitely decreasing functions:

$$\lim_{x\to 0}\frac{\alpha(x)}{\beta(x)}=\lim_{x\to 0}\frac{\sqrt{1+x}-1}{\frac{1}{2}x}=\lim_{x\to 0}\frac{1+x-1}{\frac{1}{2}x\left(\sqrt{1+x}+1\right)}=\lim_{x\to 0}\frac{2}{\sqrt{1+x}+1}=1.$$

So, as $\lim\limits_{x\to 0}\dfrac{\alpha(x)}{\beta(x)}=1$, the functions $\alpha(x)$ and $\beta(x)$ are equivalent infinitely decreasing functions provided $x\to 0$. In this case, at the values of x near 0, the value of $\beta(x)$ differs slightly from the approximate value of $\beta(x)$:

$$\alpha(x)\approx\beta(x)\Rightarrow\sqrt{1+x}-1\approx\frac{1}{2}x.$$

Solution of 3): Now, using the obtained approximate calculation formula, find the approximate value of the number $\sqrt{1672}$. For that represent $\sqrt{1672}$ as follows:

$$\sqrt{1672}\approx\sqrt{1600\cdot 1.044}=40\cdot\sqrt{1+0.044}.$$

In the formula $\sqrt{1+x}-1\approx\frac{1}{2}x$ when we take x = 0.044,we get

$$\sqrt{1+0.044}-1\approx\frac{1}{2}\cdot 0.044=0.022\Rightarrow\sqrt{1.044}\approx 1.022.$$

Taking the approximate value obtained for $\sqrt{1.044}$ into account in formula $\sqrt{1672} \approx 40 \cdot \sqrt{1+0.044}$, we get $\sqrt{1672} \approx 40 \cdot 1.022 = 40.88$.

Answer: 1) $\sqrt{106} \approx 10.3$; 2) $\sqrt{912} \approx 30.195$.

Problem 311. Replacing the infinitely decreasing functions by appropriate equivalent infinitely decreasing ones, calculate the following limits:

$$1) \lim_{x \to 0} \frac{\sqrt{1+x+x^2}-1}{\sin 2x}; \; 2) \lim_{x \to 0} \frac{\sqrt{1+x^2}-1}{1-\cos x};$$

$$3) \lim_{x \to 0} \frac{\sin 5x}{x+x^2}; \; 4) \lim_{x \to 0} \frac{\arcsin x}{arctg 6x}; \; 5) \lim_{x \to 1} \frac{\sin^2 (x-1)}{x^2-1}.$$

Solution of 1): In this section, if in the given equivalence relation (2.16) we take $\alpha(x) = x + x^2$, $n = 2$ we get,

$$\sqrt{1+x+x^2}-1 \sim \frac{x+x^2}{2}.$$

Take this into account in the given limit:

$$\lim_{x \to 0} \frac{\sqrt{1+x+x^2}-1}{\sin 2x} = \lim_{x \to 0} \frac{\frac{1}{2}\left(x+x^2\right)}{\sin 2x} = \frac{1}{4} \lim_{2x \to 0} \frac{1}{\frac{\sin 2x}{2x}} +$$

$$+\frac{1}{2} \lim_{x \to 0} \frac{x}{\frac{\sin 2x}{2x}} = \frac{1}{4} \cdot 1 + \frac{1}{2} \cdot \frac{0}{1} = \frac{1}{4}.$$

Answer: 2) 1.

Guideline: Use equivalence relations 10) and 3); 3) 5.

Guideline: Use equivalence relations 1); 4) $\frac{1}{6}$.

Guideline: Use equivalence relations 4) and 5); 5) 0.

Home tasks

Problem 312. Using the Heine definition of the limit of a function, show that,

$$\lim_{x \to 2} \frac{3x+1}{5x+4} = \frac{1}{2}.$$

Problem 313. Using the Cauchy definition of the limit of a function, prove that,

$$\lim_{x\to 1}(3x-8)=-5.$$

Problem 314. Prove that the following limits are absent:

$$1)\ \lim_{x\to\infty}\sin x\ ;\ 2)\ \lim_{x\to\infty}\cos x\,.$$

Guideline: Take in (1) two sequences of values of x in the form $x_n = \pi n$ and $x_n' = 2n\pi+\frac{\pi}{2}$, $(n=1,2,\ldots)$, use the Heine definition.

Problem 315. Find the limits:

$$1)\lim_{x\to 7}\frac{2x^2-11x-21}{x^2-9x+14};2)\lim_{x\to 1}\left(\frac{3}{1-x^3}+\frac{1}{x-1}\right);$$

$$3)\lim_{x\to\frac{\pi}{4}}\frac{\sin x-\cos x}{\cos 2x};4)\lim_{x\to 1}\frac{\sqrt[3]{x}-1}{\sqrt{x}-1}.$$

Answer $1)\frac{17}{5};2)1;3)-\frac{1}{\sqrt{2}};4)\frac{2}{3}$.

Problem 316. Find the limits:

$$1)\lim_{x\to 0}\frac{x}{\sqrt[3]{x+1}-1};2)\lim_{x\to 0}\frac{2\sqrt{x^2+x+1}-2-x}{x^2};$$

$$3)\lim_{x\to 2}\frac{\sqrt{7+2x-x^2}-\sqrt{1+x+x^2}}{2x-x^2}.$$

Answer: $1)3;2)\frac{3}{4};3)\frac{\sqrt{7}}{4}$.

Problem 317. Find the limits:

$$1)\lim_{x\to\infty}\left(\sqrt{x^4+2x^2-1}-\sqrt{x^4-2x^2-1}\right);$$

$$2)^*\lim_{x\to\infty}\left(\sqrt[3]{x^3+3x^2+4x}-\sqrt[3]{x^3-3x^2+4}\right);$$

$$3)^*\lim_{x\to\infty}x^2\left(\sqrt{x^4+x^2\sqrt{x^4+1}}-\sqrt{2x^4}\right).$$

Answer: $1)2;2)2;3)\frac{\sqrt{2}}{8}$.

Problem 318. Calculate the limits:

$$1)\lim_{x\to-1}\frac{x+1}{\sqrt[4]{x+17}-2};2)\lim_{x\to0}\frac{\sqrt[k]{1+x}-1}{x},\ (k>0\ \text{is an integer});$$

$$3)\lim_{x\to\frac{\pi}{2}}\frac{\cos x}{\sqrt[3]{(1-\sin x)^2}}.$$

Answer: 1) 32.

Guideline: Use the substitution x + 17 = t^4 ; 2) $\frac{1}{k}$.

Guideline: Use the substitution $1+x=z^k$ and the formula

$z^k-1=(z-1)\left(z^{k-1}+z^{k-2}+\ldots+z+1\right)$; 3) ∞.

Guideline: Use the substitution $x=\frac{\pi}{2}-z$ and the formula $1-\cos z=2\sin^2\frac{z}{2}$.

Problem 319. Calculate the limits:

$$1)\lim_{x\to\pi}\frac{\sin 7x}{tg3x};2)\lim_{x\to0}\frac{3\arcsin x}{4x};$$

$$3)\lim_{x\to0}\frac{\cos\alpha x-\cos\beta x}{x^2},(\alpha\neq\beta);$$

$$4)\lim_{x\to0}\left(\frac{1}{\sin x}-ctgx\right).$$

Answer: $1)\frac{7}{3};2)\frac{3}{4};3)\frac{\alpha^2-\beta^2}{2};4)0$.

Problem 320. Find the limits:

$$1)\lim_{x\to\infty}\left(\frac{x+3}{x-2}\right)^{2x+1};2)\lim_{x\to\infty}\left(\frac{x^2+5}{x^2-5}\right)^{x^2};$$

$$3)\lim_{x\to0}(\cos x)^{\frac{1}{x^2}};4)\lim_{x\to0}\left(1+tg^2\sqrt{x}\right)^{\frac{3}{x}}.$$

Answer: 1) e^{10}.

Guideline: Use (2.11); 2) e^{10}.

Guideline: Use (2.11); 3) $e^{-\frac{1}{2}}$.

Guideline: Use (2.10); 4) e^{10}.

Guideline: Use (2.10).

Problem 321. Calculate the limits:

$$1)\lim_{x\to\infty} x\left(\ln(2+x)-\ln x\right);2)\lim_{x\to 0}\frac{1}{x}\cdot\ln\sqrt{\frac{1+x}{1-x}};$$

$$3)\lim_{x\to\infty} x(a^{\frac{1}{x}}-1);4)\lim_{x\to 1}\frac{a^x-a}{x-1};5)\lim_{x\to a}\frac{\log_a x-1}{x-a};$$

$$6)\lim_{x\to 0}\frac{e^{ax}-e^{bx}}{x};7)\lim_{x\to 0}(\cos x)^{\frac{1}{\sin x}};8)\lim_{x\to 0}\frac{\ln\cos x}{x^2}.$$

Answer: 1) 2; 2) 1; 3) ln a; 4) a ln a; 5) $\frac{1}{a}\log_a e$; 6) $a-b$; 7) 1; 8) $-\frac{1}{2}$.

Problem 322. Find the limits:

$$1)\lim_{x\to 0}\left(\frac{1+tg\,x}{1+\sin x}\right)^{\frac{1}{\sin x}};2)\lim_{x\to 1}\left(1+\sin\pi x\right)^{ctg\pi x}.$$

Answer: 1) 1; 2) $\frac{1}{e}$.

Guideline: Use formula (2.17).

Problem 323. Find the represented right and left limits of the following functions:

$$1)\,f(x)=\begin{cases}-2x+3, & x\le 1,\\ 3x-5, & x>1,\end{cases}\quad \lim_{x\to 1\pm 0} f(x);$$

$$2)\,f(x)=\frac{x^2-1}{|x-1|},\ \lim_{x\to 1\pm 0} f(x);$$

$$3)\,f(x)=\frac{\sqrt{1-\cos 2x}}{x},\ \lim_{x\to 0\pm 0} f(x);$$

$$4)\,f(x)=3+\frac{1}{1+7^{\frac{1}{1-x}}},\ \lim_{x\to 1\pm 0} f(x).$$

Answer: 1) $\lim\limits_{x\to 1-0} f(x)=1,\ \lim\limits_{x\to 1+0} f(x)=-2;$

2) $\lim\limits_{x\to 1-0} f(x)=-2,\ \lim\limits_{x\to 1+0} f(x)=2;$

3) $\lim\limits_{x\to 0-0} f(x)=-\sqrt{2},\ \lim\limits_{x\to 0+0} f(x)=\sqrt{2};$

4) $\lim\limits_{x\to 1-0} f(x)=3,\ \lim\limits_{x\to 1+0} f(x)=4.$

Problem 324. Show that the following functions are infinitely increasing at the given points:

1) at the point $f(x)=\dfrac{3x-12}{2x^2+7},\ x=4;$

2) at the point $f(x)=x\sin\dfrac{1}{x},\ x=0.$

Problem 325. Compare the infinitely decreasing functions:

1) $\alpha(x)=e^{x-1}-1, \beta(x)=2x-2,$ provided $x\to 1;$

2) $\alpha(x)=\sin x^3, \beta(x)=x^2,$ provided $x\to 0;$

3) $\alpha(x)=\lg(1+x), \beta(x)=x,$ provided $x\to 0.$

Answer: 1) $\alpha \sim \beta$; 2) $\alpha = o(\beta)$; 3) are of the same order.

Problem 326. Determine approximate order of the infinitely decreasing function $\alpha(x)$ with respect to the infinitely decreasing $\beta(x)$:

1) $\alpha(x)=e^{x^3}-1, \beta(x)=x,$ provided $x\to 0;$

2) $\alpha(x)=\ln\left(1+\sqrt[3]{x^2}\right), \beta(x)=\sqrt[3]{x},$ provided $x\to 0.$

Answer: 1) $\alpha(x)$ is a third-order infinitely decreasing function with respect to $\beta(x)$; 2) $\alpha(x)$ is a second-order infinitely decreasing function with respect to $\beta(x)$.

Problem 327. Prove that as $x\to 0$ the functions $\sqrt[n]{1+x}-1$ and $\dfrac{x}{n}$ (n is a natural number) are equivalent infinitely decreasing. Using this equivalence, find the approximate value of the number $\sqrt[4]{1024}$.

Answer: 10.03.

Problem 328. Using the approximate equivalence conditions, find the following limits:

$$1)\ \lim_{x\to 0}\frac{\sqrt[3]{1+\sin x}-1}{tg\,x};\ 2)\ \lim_{x\to 0}\frac{\left(\sqrt[n]{1+tg\,x}-1\right)\left(\sqrt{1+x}-1\right)}{2x\sin x};$$

$$3)\ \lim_{x\to 2}\frac{arctg(2-x)+\sin(x-2)^2}{x^2-4};\ 4)\ \lim_{x\to 0}\frac{\sqrt[4]{1+x^2}+x^3-1}{\ln\cos x};$$

$$5)\ \lim_{x\to 0}\frac{x^3\sqrt[10]{x}\cos x+\sin^3 x}{1-\sqrt{1+x^3}}.$$

Answer: $1)\frac{1}{3};2)\frac{1}{4n};\ 3)-\frac{1}{4};\ 4)-\frac{1}{2};\ 5)-2.$

2.4 CONTINUITY OF FUNCTIONS. DISCONTINUITY POINTS AND THEIR CLASSIFICATION

Assume that the function $y=f(x)$ is determined in the set X and a is the limit point of X.

Definition. When for the function $f(x)$ the condition

$$\lim_{x\to a} f(x)=f(a) \tag{2.18}$$

is satisfied, the function $f(x)$ is called a continuous function at the point a.

The number $f(x)$ is said to be an eigenvalue of the function $f(x)$ at the point a. Therefore, definition of the function continuous at the point may be expressed as follows.

If a function has a limit value at the given point and equals its eigenvalue at that point, then this function is called a continuous function at this point.

Using definition of the function's limit in the terms of "$\varepsilon-\delta$," based on (2.18) we can give the following definition to the function continuous at a point.

Definition. If for any number $\varepsilon>0$ one can find a number $\delta>0$ such that in all the values of x satisfying the inequality $|x-a|<\delta$, $|f(x)-f(a)|<\varepsilon$, then the function $f(x)$ is called a continuous function at the point a.

This is definition of continuity in the terms of "$\varepsilon-\delta$."

If we take $x = a + \Delta x$, it is clear that provided $x\to a$, $\Delta x\to 0$. Therefore, we can write (1) as follows:

$$\lim_{\Delta x\to 0} f(a+\Delta x) = f(a) \Rightarrow$$
$$\Rightarrow \lim_{\Delta x\to 0}[f(a+\Delta x)-f(a)] = 0 \Rightarrow \lim_{\Delta x\to 0}\Delta f = 0. \quad (2.19)$$

Equality (2.19) expresses the definition of a continuous function "in the sense of increment."

Definition. If the equality $\lim_{x\to a-0} f(x) = f(a)$ $\left(\lim_{x\to a+0} f(x) = f(a)\right)$ is satisfied, the function $f(x)$ is called a right (left) continuous function at the point a.

For the function to be continuous at the given point it is necessary and sufficient this function to be both right and left continuous at this point.

A function continuous at any point of the set X is said to be a continuous function in this set.

The function $f(x)$ is called a continuous function on the interval $[a, b]$ if this function is continuous at all internal points of the interval $[a, b]$, right continuous at the point a, left continuous at the point b.

Show some properties of functions continuous at a point. The function continuous at a point is bounded in certain neighborhood of this point.

If the function $f(x)$ is continuous and $f(a) \neq 0$, then the point a has a neighborhood such that for all x points from this neighborhood, the sign of $f(x)$ is the same with the sign of $f(a)$.

If the functions $f(x)$ and $g(x)$ are continuous at the point a, then the functions, $f(x)\pm g(x), f(x)\cdot g(x), \frac{f(x)}{g(x)}, (g(a)\neq 0)$ are also continuous at this a point.

If the function $u = \varphi(x)$ is continuous at the point x_0, the function $y = f(u)$ at the point $u_0 = \varphi(x_0)$, then the complex function $u = f(\varphi(x))$ is continuous at the point x_0.

The function $f(x)$ determined in the set X is said to be a regularly continuous function in this set if for any number $\varepsilon > 0$ it is possible to take a number $\delta(\varepsilon) > 0$ dependent only on the chosen ε such that for arbitrary x' and x'' values of x from X and satisfying the condition $|x'-x''| < \delta$, $|f(x')-f(x'')| < \varepsilon$.

We give some properties of a function continuous on an interval.

Theorem 1. (Weierstrass's first theorem). A function continuous on an interval is bounded on this interval.

Theorem 2. (Weierstrass's second theorem). A function continuous on an interval takes its exact lower and exact upper bounds.

Theorem 3. If a function continuous on the interval $[a, b]$ takes different values at the edges of this interval, then this function vanishes at least at one point inside of the interval.

Theorem 4. If the function $f(x)$ continuous on the interval $[a, b]$ takes the values $f(a) = A, f(b) = B, (A \neq B)$ at the edges of this interval, then the function $f(x)$ at least at one point takes a value equal to any arbitrary number C located between A and B: $(f(\xi) = C, \xi \in (a, b))$.

Theorem 5. A function continuous on the interval is regularly continuous on this interval.

If for the function $f(x)$ the condition (2.18) is not satisfied at the point a, then the point a is called a discontinuity point of the function $f(x)$.

(1) May not be satisfied by many reasons. There may be so that the function $f(x)$ has no limit at the point a, $f(x)$ be not determined at the point a, the point a has a limit value and the function is determined at the point a, but $\lim_{x\to a} f(x) \neq f(a)$. Depending on these cases the following classification of discontinuity points is given.

1) If the function $f(x)$ has a limit value at the point a but the function $f(x)$ was not determined at the point a or the eigenvalue of $f(a)$ at this point is not equal to the limit value at this point, then the point a is said to be removable discontinuity point of $f(x)$.

For example, as for the function $f(x) = \begin{cases} \dfrac{\sin x}{x}, x \neq 0, \\ 2, \quad x = 0 \end{cases}$

$\lim_{x\to 0} f(x) = \lim_{x\to 0} \dfrac{\sin x}{x} = 1,\ f(0) = 2,\ \lim_{x\to 0} f(x) \neq f(0)$, then $x = 0$ is removable discontinuity point of this function. Indeed if we take $f(0) = 1$, that is, accept

$$f(x) = \begin{cases} \dfrac{\sin x}{x}, x \neq 0 \\ 1, \quad x = 0 \end{cases},$$

the function $f(x)$ is discontinuous at the point $x = 0$.

2) If the function $f(x)$ at the point a has both right and left limits and is finite, but is not equal to each other, then the point a is called the first kind discontinuity point of the function $f(x)$. The number

$\lim\limits_{x\to a+0} f(x) - \lim\limits_{x\to a-0} f(x) = \Delta f(a)$ is called a jump of the function $f(x)$ at the point a.

For example, as for the function $sign x = \begin{cases} 1, x > 0, \\ 0, x = 0, \\ -1, x < 0 \end{cases}$ $\lim\limits_{x\to 0-0} sign x = -1$, $\lim\limits_{x\to 0+0} sign x = 1$, then $x = 0$ is the first kind discontinuity point for this function and

$$\Delta sign(0) = 1 - (-1) = 2 .$$

3) If the function has not at least one of the left and right limits at the point a or is not finite, then the point a is called the second discontinuity point of the function $f(x)$.

For example, the function $f(x) = \sin\frac{1}{x}$ has either right or left limit at the point $x = 0$. Therefore, $x = 0$ is the second kind discontinuity point for this function.

Problems to be solved in auditorium

Problem 329. Using the definition $\lim\limits_{x\to a} f(x) = f(a)$ of continuity of functions, show that the following functions are continuous at the given point:

1) at the point $f(x) = (x+1) arctg\, x,\ x = 1$;

2) $f(x) = \begin{cases} -\frac{1}{2}x^2, x \le 2, \\ x, \quad x > 2, \end{cases}$ at the point $x = 3$;

3) $f(x) = \begin{cases} 2\sqrt{x}, 0 \le x \le 1, \\ 4 - 2x, 1 < x < \infty, \end{cases}$ at the point $x = 3$.

Problem 330. Using the definition of continuity of the function in terms of "$\varepsilon - \delta$," show that the function $f(x) = \sqrt{x}$ is continuous at any $a > 0$ point and right continuous at the point $a = 0$.

Solution. At first, we note that for any numbers $x > 0$, $a > 0$ we can write

$$0 \le \left|\sqrt{x} - \sqrt{a}\right| = \frac{|x - a|}{\sqrt{x} + \sqrt{a}} \le \frac{|x - a|}{\sqrt{a}}$$

From the last relation, we get that if for any $\varepsilon > 0$ we take a number δ satisfying the condition $0 < \delta < \sqrt{a}\varepsilon$, for $0 < |x-a| < \delta$ we get

$$|f(x)-f(a)| = |\sqrt{x}-\sqrt{a}| \le \frac{|x-a|}{\sqrt{a}} < \frac{\delta}{\sqrt{a}} < \frac{\sqrt{a}\varepsilon}{\sqrt{a}} = \varepsilon \Rightarrow |f(x)-f(a)| < \varepsilon.$$

This shows that the function $f(x) = \sqrt{x}$ is continuous at any point $a > 0$.

Now show that the function $f(x) = \sqrt{x}$ is right continuous at the point $a > 0$. For that, we should show that for any number $\varepsilon > 0$ we can find a number $\delta > 0$ such that at all values of x satisfying the inequality $0 \le x - 0 < \delta \Rightarrow 0 \le x < \delta$ $|f(x)-f(0)| = |\sqrt{x}-\sqrt{0}| < \varepsilon \Rightarrow \sqrt{x} < \varepsilon$. It is clear that if for the chosen ε we take $\delta = \varepsilon^2$, for $0 \le x < \delta$,

$$\sqrt{x} < \sqrt{\delta} \Rightarrow \sqrt{x} < \sqrt{\varepsilon^2} \Rightarrow \sqrt{x} < \varepsilon \Rightarrow |f(x)-f(0)| < \varepsilon,$$

that is, function $f(x) = \sqrt{x}$ is right continuous at the point $a = 0$.

Problem 331. Using the definition of continuity of a function "in increment sense," prove that the function $f(x) = x^3$ is continuous at any a point.

Solution. Let us give any small increment Δx to x at the point a and calculate appropriate increment of the function:

$$\Delta f = f(a+\Delta x) - f(a) = (a+\Delta x)^3 - a^3 = a^3 + 3a^2\Delta x + 3a(\Delta x)^2 +$$
$$+(\Delta x)^3 - a^3 = \Delta x\left(3a^2 + 3a\Delta x + (\Delta x)^2\right).$$

As $\lim\limits_{\Delta x \to 0} \Delta f = 0$, the function $f(x) = x^3$ is continuous at any a point.

Problem 332. Prove that the below-given function $f(x)$ is continuous at each point contained in the domain of definition:

$$1)\, y = x^2;\, 2)\, y = \frac{1}{x};\, 3)\, y = |x|;\, 4)\, y = \sqrt[3]{x}.$$

Problem 333. Near the point a with points $x \ne a$ the function $f(x)$ is determined by the following formula:

$$1) f(x)=\frac{x^2-1}{x+1}, x\neq -1;\ 2) f(x)=\frac{\sqrt{1+x}-1}{x}, x\neq 0;$$
$$3) f(x)=x\,ctg\,x, x\neq 0.$$

Determine the function $f(x)$ at the point a such that the obtained function be continuous at the point a.

Answer $1) f(-1)=-2;\ 2) f(0)=\frac{1}{2};3) f(0)=1$.

Guideline: Calculate $\lim\limits_{x\to a} f(x)$ and take

$$f(a)=\lim_{x\to a} f(x).$$

Problem 334. Prove that $x = 3$ is the first kind discontinuity point for the function $f(x)=arctg\frac{1}{x-3}$.

Solution. At the point $x = 3$ of this function calculate the right and left limits:

$$\lim_{x\to 3+0} f(x)=\lim_{\substack{x\to 3\\ x>3}} arctg\frac{1}{x-3}=\frac{\pi}{2},$$

$$\lim_{x\to 3-0} f(x)=\lim_{\substack{x\to 3\\ x<3}} arctg\frac{1}{x-3}=-\frac{\pi}{2}.$$

As $\lim\limits_{x\to 3+0} f(x)\neq \lim\limits_{x\to 3-0} f(x)$, $x=3$ is the first discontinuity point for this function.

Problem 335. Study continuity of the functions:

$$1) f(x)=\frac{|3x-5|}{3x-5};2) f(x)=\begin{cases}2x+5, -\infty<x<-1\\ \frac{1}{x}, \quad -1\le x<\infty;\end{cases}$$

$$3) f(x)=\begin{cases}x+4, -\infty<x<-1,\\ x^2+2, -1\le x<1,\\ 2x, \quad 1\le x<+\infty;\end{cases}\quad 4) f(x)=\begin{cases}x+2, -\infty<x\le -1,\\ x^2+1, -1<x\le 1,\\ -x+3,\ 1<x<+\infty;\end{cases}$$

$$5)\, f(x)=\begin{cases}-x, & -\infty<x\le 0,\\ -(x-1)^2, & 0<x<2,\\ x-3, & 2\le x<+\infty;\end{cases}\quad 6)\, f(x)=\begin{cases}\cos x, & -\frac{\pi}{2}\le x<\frac{\pi}{4},\\ x-\frac{\pi^2}{16}, & \frac{\pi}{4}<x\le\pi;\end{cases}$$

$$7)\, f(x)=\begin{cases}-\frac{1}{x}, & -\infty<x<0,\\ 1, & 0\le x<1,\\ x, & 1\le x\le 2,\\ 3, & 2<x\le 3.\end{cases}$$

Answer: 1) $x=\frac{5}{3}$ is the first kind discontinuity point; 2) $x=-1$ is the first kind, $x=0$ is the second kind discontinuity point; 3) $x=1$ is the first kind discontinuity point; 4) $x=-1$ is the first kind discontinuity point; 5) $x=0$ is the first kind discontinuity point; 6) $x=\frac{\pi}{4}$ is the first kind discontinuity point; 7) at the point $x=0$ the function is left discontinuous, right continuous. $x=0$ is the second kind, $x=2$ is the first kind discontinuity point.

Problem 336. Prove that every point of the number axis is a discontinuity point for the Dirichlet function.

Solution. Let us take any real number a. Two cases are possible:

1) **Case 1.** a is a rational number. Then, $D(a)=1$. At any small vicinity of the point a there exist irrational numbers x and for these points $D(x)=0$. Therefore the modulus of increment for the function at the point a will be

$$|\Delta D|=|D(x)-D(a)|=|0-1|=1.$$

2) **Case 2.** a is an irrational number. Then $D(a)=0$. In any small vicinity of the point a there exist x rational numbers and for these points $D(a)=1$. The modulus of increment of the function at the point a will be

$$|\Delta D|=|D(x)-D(a)|=|1-0|=1.$$

As is seen, when a is any real number, the increment ΔD of the Dirichlet function at the point a is not infinitely decreasing provided $\Delta x \to 0$ $(x \to a)$. So, the Dirichlet function is discontinuous at each point.

Problem 337. Study that the functions are continuous:

1) $y = \cos x^n$, (n is a natural number);
2) $y = \cos \lg x$.

Solution of 1): Take $u = x^n$, $y = \cos u$ and consider the given function as a complex function. The function $y = \cos u$ at any u point, the function $u = x^n$ at any value of x are continuous (when n is a natural number). Therefore, the complex number $u = \cos x^n$ is continuous at any point of the number axis.

Answer: 2) is continuous on the interval $(0, +\infty)$.

Problem 338. Prove that the function $\sin x - x + 1 = 0$ has at least one real root.

Solution. At first, note that the function $f(x) = \sin x - x + 1$ is continuous on the whole of the axis. Furthermore, as $f(0) = 1 > 0$, $f\left(\frac{3\pi}{2}\right) = -\frac{3\pi}{2} < 0$, on the interval $\left[0, \frac{3\pi}{2}\right]$ this function gets values with different signs. By theorem 3, inside of the interval $\left[0, \frac{3\pi}{2}\right]$ there is at least one point that $f(x) = 0 \Rightarrow \sin x - x + 1 = 0$.

Problem 339. Determine if the following equations have roots in the indicated interval:

$$1)\, x^5 - 18x + 2 = 0,\ [-1;1];$$

$$2)\, f(x) = \begin{cases} x^2 + 2, & -2 \le x < 0, \\ -\left(x^2 + 2\right), & 0 \le x \le 2, \end{cases} \quad [-2;2].$$

Answer: 1) has; 2) has not.

Problem 340. Show that the function $f(x) = \frac{1}{4}x^3 - \sin \pi x + 3$ gets the value equal to $2\frac{1}{3}$.

Guideline: Consider the function $f(x)$ on the interval $[-2;2]$ and use theorem 4.

Problem 341. Show that the following functions are regularly continuous in the indicated intervals:

$$1)\ f(x)=\frac{x}{4-x^2},[-1;1];\ 2)\ f(x)=\sin x,(-\infty;+\infty).$$

Guideline: In 2) Show the difference in the form of product $\sin x' - \sin x''$ and use the relation $\left|\sin\frac{x'-x''}{2}\right| \le \frac{|x'-x''|}{2}$.

Home tasks

Problem 342. Show that the below-given function $f(x)$ is continuous at every point contained in domain of definition:

$$1)\ y=2x-1;2)\ y=ax+b;3)\ y=\sqrt{x};4)\ y=x^3.$$

Problem 343. The function $f(x)$ is determined at the points $x \neq 0$ in the vicinity of the point $x = 0$ by the following formula:

$$1)\ f(x)=1-x\sin\frac{1}{x},x\neq 0;2)\ f(x)=\frac{1-\cos x}{x^2},x\neq 0;$$

$$3)\ f(x)=\frac{(1+x)^n-1}{x},x\neq 0,$$

(n is a natural number). Determine the function $f(x)$ at the point $x = 0$ so that the obtained function be continuous at the point $x = 0$.

Answer: $1)\ f(0)=1;2)\ f(0)=\frac{1}{2};3)\ f(0)=n$.

Problem 344. Show that the function $f(x) = [x]$ that indicates the largest integer less than x is continuous at every entire value of the argument x.

Problem 345. Study the continuity of functions. Find the jumps at the discontinuity points:

$$1)\ y=\frac{|x+2|}{x+2};2)\ y=\frac{1}{x^2-4};3)\ y=\begin{cases}\frac{1}{x-1},x<0,\\ (x+1)^2,\ 0\le x\le 2,\\ 1-x,x>2.\end{cases}$$

Answer: 1) $x = -2$ is the first kind discontinuity point, $\Delta y(-2) = 2$; 2) $x = \pm 2$ are the second kind discontinuity points; 3) $x = 0$ and $x = 2$ are the first kind discontinuity points, $\Delta y(0) = 2$, $\Delta y(2) = -10$.

Problem 346. At which value of a the function $f(x)$ is continuous:

$$1)\ f(x) = \begin{cases} \dfrac{(1+x)^n - 1}{x}, & x \neq 0, \\ a, & x = 0; \end{cases} \quad (n \text{ is a natural number})$$

$$2)\ f(x) = \begin{cases} e^{-\frac{1}{x^2}}, x \neq 0, \\ a,\ x = 0; \end{cases} \quad 3)\ f(x) = \begin{cases} (1+x)^{\frac{1}{x}}, x \neq 0, \\ a, \quad x = 0; \end{cases}$$

$$4)\ f(x) = \begin{cases} \dfrac{c^x - 1}{x}, x \neq 0, (c > 0) \\ a, \quad x = 0. \end{cases}$$

Answer: 1) $a = n$; 2) $a = 0$; 3) $a = e$; 4) $a = \text{h}\ c$

Problem 347*. Prove that the function

$$f(x) = \begin{cases} x \quad \text{when} \quad x \quad \text{is a rational number,} \\ 0 \quad \text{when} \quad x \quad \text{is an irrational number} \end{cases}$$

is continuous at the point $x = 0$, is discontinuous at all remaining points.

Problem 348. Study the continuity of the function $f(x) = \dfrac{1}{x^2 - 7x + 6}$ on the following intervals:

$$1)[2;5]; \quad 2)[4;10]; \quad 3)[0;7].$$

Answer: 1) is continuous; 2) at the point $x = 6$ has second kind discontinuity; 3) at the points $x = 1$ and $x = 6$ has second kind discontinuity.

Problem 349. Show that the equation $x^5 - 3x - 1 = 0$ has at least one real root between 1 and 2.

Problem 350. Prove that the function $f(x) = \dfrac{x^2}{2} + +\sin\dfrac{\pi x}{2} + 1$ takes a value equal to 1.

Guideline: One can consider the function $f(x)$ on the interval $[-1;1]$ and use theorem 4.

Problem 351*. Prove that the function $f(x)=\sin\frac{1}{x}$ is not regularly continuous on the interval (0; 1).

Guideline: Take $x'=\frac{2}{(4k+1)\pi}$, $x''=\frac{2}{(4k+3)\pi}$ though at rather great values of k for any $\delta > 0$ the condition $|x'-x''|<\delta$ is satisfied, use that $\left|\sin\frac{1}{x'}-\sin\frac{1}{x''}\right|=2$.

Problem 352. Show that the below-given functions are regularly continuous on the given intervals:

$$1)\ f(x)=\sqrt[3]{x},[0;2];\ 2)\ f(x)=\frac{1}{x},[a,+\infty),(a>0)$$

Guideline: In 2) use that $\left|\frac{1}{x'}-\frac{1}{x''}\right|\le\frac{1}{a^2}|x'-x''|$ for any x', $x'' \in [a, +\infty]$, $(a>0)$.

KEYWORDS

- **function**
- **domain of definition**
- **set of values**
- **even function**
- **odd function**
- **periodic function**
- **period of a function**
- **numerical sequence**
- **convergent numerical sequence**
- **limit of a function**
- **infinitely decreasing functions**
- **continuous functions**
- **discontinuity points**

CHAPTER 3

DIFFERENTIAL CALCULUS OF A FUNCTION OF ONE VARIABLE

CONTENTS

ABSTRACT

In this chapter, we give brief theoretical materials on definition of the derivative of an one-variable function and its differential, their applications, main theorems of differential calculus, Taylor and Maclaurin formulae and their applications, indefinitenesses, and their calculation rules, and 64 problems.

3.1 DERIVATIVE AND ITS CALCULATING RULES

Assume that the function $y = f(x)$ was determined in certain neighborhood of the indicated point x. At this point to the argument, we give an increment Δx such that $x + \Delta x$ be contained in the neighborhood of x. Let us consider the following ratio:

$$\frac{\Delta y}{\Delta x} = \frac{f(x+\Delta x) - f(x)}{\Delta x}. \tag{3.1}$$

When the x value of the argument is in the fixed point, ratio (3.1) is a Δx dependent function and is determined in certain neighborhood of the point $\Delta x = 0$ (except the point $\Delta x = 0$ itself).

Definition. Provided $\Delta x \to 0$, if ratio (3.1) has a finite limit, this limit is said to be the derivative of the function $y = f(x)$ at the point x and is denoted by one of the following symbols

$$y', \; f'(x), \; \frac{dy}{dx}, \; \frac{df(x)}{dx}.$$

By definition, we can write

$$f'(x) = \lim_{\Delta x \to 0} \frac{f(x+\Delta x) - f(x)}{\Delta x}. \tag{3.2}$$

If relation (3.1) has finite right (left) limit at the point $\Delta x = 0$, this limit is said to be the right (left) limit of the function $f(x)$ and is denoted as $f'_+(x)$ $(f'_-(x))$. For the function

$$f'_+(x) = \lim_{\substack{\Delta x \to 0 \\ \Delta x > 0}} \frac{f(x+\Delta x) - f(x)}{\Delta x}, \quad f'_-(x) = \lim_{\substack{\Delta x \to 0 \\ \Delta x < 0}} \frac{f(x+\Delta x) - f(x)}{\Delta x},$$

to have a derivative at the point x, the existence and equality of the finite left and right derivatives of this function is a necessary and sufficient condition.

There may happen so that the function has both a left and right derivative at the given point, but has not a derivative at this point.

For example, at the point $x = 0$ the function $f(x) = |x|$ has both left and right derivatives:

$$f'_{-}(0) = \lim_{\substack{\Delta x \to 0 \\ \Delta x < 0}} \frac{|0 + \Delta x| - |0|}{\Delta x} = \lim_{\substack{\Delta x \to 0 \\ \Delta x < 0}} \frac{|\Delta x|}{\Delta x} = \lim_{\Delta x \to 0} \frac{-\Delta x}{\Delta x} = -\lim_{\Delta x \to 0} 1 = -1,$$

$$f'_{+}(0) = \lim_{\substack{\Delta x \to 0 \\ \Delta x > 0}} \frac{|\Delta x|}{\Delta x} = \lim_{\Delta x \to 0} \frac{\Delta x}{\Delta x} = \lim_{\Delta x \to 0} 1 = 1.$$

As $f'_{-}(0) \neq f'_{+}(0)$, the function $f(x) = |x|$ has no derivative at the point $x = 0$.

Definition. If the Δy function argument of the function $y = f(x)$ corresponding to the Δx argument increment at the point x may be shown in the form

$$\Delta y = A \cdot \Delta x + \alpha(\Delta x) \cdot \Delta x, \tag{3.3}$$

the function $f(x)$ is called a differentiable function at the point x. Here A is a certain number, $\alpha(\Delta x)$ is an infinitely decreasing function of Δx provided $\Delta x \to 0$:

$$\lim_{\Delta x \to 0} \alpha(\Delta x) = 0.$$

The number A equals the value of $f(x)$ at the fixed point x.

Theorem. For the function to be differentiable at the given point it is necessary and sufficient for this function to have a finite derivative at this point.

Based on this theorem, the function that has a finite derivative at this point is called a function differentiable at this point. The operation of finding the derivative of the function is called differentiation of a function.

We show the main rules of differentiation:

$$1)\,(c)'=0,\ c=const;\ 2)\,\big(c\cdot u(x)\big)'=c\,u'(x);$$

$$3)\,\big(u(x)\pm v(x)\big)'=u'(x)\pm v'(x);$$

$$4)\left(\frac{u(x)}{v(x)}\right)'=\frac{u'(x)v(x)-u(x)v'(x)}{v^2(x)},\ v(x)\neq 0.$$

Theorem (differentiation of complex functions). When the function $u=\varphi(x)$ at the point x_0 and the function $y=f(u)$ at the point $u_0=\varphi(x_0)$ are differentiable, the complex function $y=f(\varphi(x))$ is differentiable at the point x_0 and its derivative is calculated by the formula

$$\big(f(\phi(x_0))\big)'=f'(u_0)\cdot\phi'(x_0)\,. \tag{3.4}$$

Theorem (differentiation of inverse function). If the function $y=f(x)$ is differentiable at the point x_0, and $f'(x_0)\neq 0$, then the inverse function $x=f^{-1}(y)$ of $f(x)$ is differentiable at the point $y_0(y_0=\varphi(x_0))$ and for its derivative the formula

$$\big(f^{-1}(y_0)\big)'=\frac{1}{f'(x_0)} \tag{3.5}$$

is valid.

Below we give a table of derivatives of main elementary functions. In the first formula given for every elementary function it is considered that x is an independent variable, in the second formula u is a function of a certain variable.

$$1)\ \left(x^n\right)'=nx^{n-1},\ \left(u^n\right)'=n\cdot u^{n-1}\cdot u';$$

$$2)\ \left(a^x\right)'=a^x\ln a,\ \left(a^u\right)'=a^u\cdot\ln a\cdot u',\ a>0;$$

$$\left(e^x\right)'=e^x,\ \left(e^u\right)'=e^u\cdot u';$$

$$3)\ (\log_a x)'=\frac{1}{x}\cdot\log_a e,\ (\log_a u)'=\frac{u'}{u}\log_a e,\ 0<a\neq 1;$$

$$(\ln x)'=\frac{1}{x},\ \ (\ln u)'=\frac{u'}{u};$$

4) $(\sin x)' = \cos x$, $(\sin u)' = u' \cdot \cos u$;

5) $(\cos x)' = -\sin x$, $(\cos u)' = -u' \cdot \sin u$;

6) $(tg\, x)' = \dfrac{1}{\cos^2 x}$, $(tg\, u)' = \dfrac{u'}{\cos^2 u}$;

7) $(ctg\, x)' = -\dfrac{1}{\sin^2 x}$, $(ctg\, u)' = -\dfrac{u'}{\sin^2 u}$;

8) $(\arcsin x)' = \dfrac{1}{\sqrt{1-x^2}}$, $(\arcsin u)' = \dfrac{u'}{\sqrt{1-u^2}}$;

9) $(\arccos x)' = -\dfrac{1}{\sqrt{1-x^2}}$, $(\arccos u)' = -\dfrac{u'}{\sqrt{1-u^2}}$;

10) $(arctg\, x)' = \dfrac{1}{1+x^2}$, $(arctg\, u)' = \dfrac{u'}{1+u^2}$;

11) $(arcctg\, x)' = -\dfrac{1}{1+x^2}$, $(arcctg\, u)' = -\dfrac{u'}{1+u^2}$;

12) $(sh x)' = ch x$, $(sh u)' = u' \cdot ch u$;

13) $(ch x)' = sh x$, $(ch u)' = u' \cdot sh u$.

$f'(x)$ is called a first-order derivative of $f(x)$. If the function $f'(x)$ has a derivative, this derivative is called a second-order derivative of $f(x)$ and is denoted by one of the symbols:

$$f''(x),\ f^{(2)}(x),\ y'',\ y^{(2)},\ \frac{d^2 y}{dx^2},\ \frac{d^2 f(x)}{dx^2}.$$

By this definition $f''(x) = \left(f'(x)\right)'$. For example, $(\sin x)'' = ((\sin x)')' = (\cos x)' = -\sin x$. The more higher order derivatives are determined in the same way:

$$y^{(n)} = \left(y^{(n-1)}\right)'.$$

Below we give formulas for calculating the n-th order derivatives of some elementary functions:

1) $\left(x^m\right)^{(n)} = m(m-1)\ldots(m-n+1)x^{m-n}$;

2) $\left(a^x\right)^{(n)} = a^x \ln^n a$, $(a > 0)$;

$$3)\ (\ln x)^{(n)} = (-1)^{n-1} \cdot \frac{(n-1)!}{x^n},\ (x > 0);$$

$$4)\ (\sin x)^{(n)} = \sin\left(x + n \cdot \frac{\pi}{2}\right);$$

$$5)\ (\cos x)^{(n)} = \cos\left(x + n \cdot \frac{\pi}{2}\right).$$

If the function $y = y(x)$ is implicitly given by the equation

$$F(x,\ y(x)) = 0, \tag{3.6}$$

then for finding y' we should consider the left side of eq. 3.6 as a complex function, differentiate (3.6) with respect to x, and solve the obtained equation with respect to u'.

Suppose that a certain function was given in the parametric form by the equations.

$$x = \varphi(t),\ y = \psi(t),\ \alpha < t < \beta. \tag{3.7}$$

If the functions $\varphi(t)$, $\psi(t)$ are differentiable and, $\varphi'(t) \neq 0$ then (3.7) determine a differentiable function $y(x)$ and its $y'(x)$ derivative is calculated by the formula

$$\frac{dy}{dx} = \frac{y_t'}{x_t'} = \frac{\psi'(t)}{\varphi'(t)}. \tag{3.8}$$

The derivative of natural logarithm of the function $y = f(x)$ is said a logarithmic derivative of this function. For the logarithmic derivative the following formula is valid.

$$(\ln y)' = \frac{y'}{y} \Rightarrow y' = (\ln y)' \cdot y, \tag{3.9}$$

We can show that the following formula is valid:

$$\begin{aligned}\left(u(x)^{v(x)}\right)' &= u(x)^{v(x)} \cdot \ln u(x) \cdot v'(x) + \\ &+ v'(x) \cdot u(x)^{v(x)-1} \cdot u'(x),\ (u(x) > 0).\end{aligned} \tag{3.10}$$

Problems to be solved in auditorium

Problem 353. By definition of derivative, find the derivatives of the following functions:

$$1)\, y = x^3;\ 2)\, y = \sin x;\ 3)\, y = tg\, x;\ 4)\, y = \ln x\,.$$

Solution of 3):

$$y' = (tg\, x)' = \lim_{\Delta x \to 0} \frac{\Delta y}{\Delta x} = \lim_{\Delta x \to 0} \frac{tg\,(x+\Delta x) - tg\, x}{\Delta x} =$$

$$= \lim_{\Delta x \to 0} \frac{\dfrac{\sin(x+\Delta x - x)}{\cos(x+\Delta x)\cos x}}{\Delta x} = \lim_{\Delta x \to 0} \frac{\sin \Delta x}{\Delta x \cos(x+\Delta x)\cos x} =$$

$$= \frac{1}{\cos x} \cdot \lim_{\Delta x \to 0} \frac{\sin \Delta x}{\Delta x} \cdot \lim_{\Delta x \to 0} \frac{1}{\cos(x+\Delta x)} = \frac{1}{\cos x} \cdot 1 \cdot \frac{1}{\cos x} = \frac{1}{\cos^2 x}$$

So, $(tg\, x)' = \dfrac{1}{\cos^2 x}$.

Answer: $1)\, 3x^2;\ 2)\cos x;\ 3)\dfrac{1}{x}$.

Problem 354. Find the solutions of the following functions:

$$1)\, y = \frac{x^3}{3} - 2x^2 + 4x - 5;\ 2)\, y = x + 2\sqrt{x};$$

$$3)\, y = \frac{1}{x} + \frac{1}{x^2} + \frac{1}{x^3};\ 4)\, y = 6\sqrt[3]{x} - 4\sqrt[4]{x};$$

$$5)\, y = x - \sin x;\ 6)\, y = x^2 ctg\, x;$$

$$7)\, y = \frac{\cos x}{1 - \sin x};\ 8)\, y = x\, arc\sin x;$$

$$9)\, y = \sqrt{x}\, arctg\, x;\ 10)\, y = x^2 \log_3 x;$$

$$11)\, y = x^n \ln x;\ 12)\, y = 2^x;\ 13)\, y = x \cdot 10^x;$$

$$14)\, y = e^x \cos x.$$

Solution of 3): At first let us write the given function in the following form:

$$y=\frac{1}{x}+\frac{1}{x^2}+\frac{1}{x^3}=x^{-1}+x^{-2}+x^{-3}.$$

Using the rule for obtaining a derivative from the sum and the formula $\left(x^n\right)'=nx^{n-1}$, we get:

$$y'=\left(x^{-1}\right)'+\left(x^{-2}\right)'+\left(x^{-3}\right)'=$$

$$=-1x^{-2}-2x^{-3}-3x^{-4}=-\frac{1}{x^2}-\frac{2}{x^3}-\frac{3}{x^4}.$$

Solution of 4):

$$y'=\left(6\sqrt[3]{x}-4\sqrt[4]{x}\right)'=\left(6x^{\frac{1}{3}}-4x^{\frac{1}{4}}\right)'=$$

$$=6\cdot\frac{1}{3}x^{-\frac{2}{3}}-4\cdot\frac{1}{4}x^{-\frac{3}{4}}=\frac{2}{\sqrt[3]{x^2}}-\frac{1}{\sqrt[4]{x^3}}.$$

Solution of 6): Using the rule for obtaining a derivative from the product and the formulas $\left(x^n\right)'=nx^{n-1}$, $(ctg\,x)'=-\dfrac{1}{\sin^2 x}$, we get:

$$y'=\left(x^2 ctg\,x\right)'=\left(x^2\right)'ctg\,x+x^2\cdot\left(ctg\,x\right)'=$$

$$=2x\,ctg\,x+x^2\cdot\left(-\frac{1}{\sin^2 x}\right)=$$

$$=\frac{2x\sin x\cos x-x^2}{\sin^2 x}=\frac{x(\sin 2x-x)}{\sin^2 x}.$$

Solution of 7): Using the rule for obtaining a derivative from the fraction, we can write:

$$y'=\left(\frac{\cos x}{1-\sin x}\right)'=\frac{(\cos x)'(1-\sin x)-\cos x\cdot(1-\sin x)'}{(1-\sin x)^2}=$$

$$=\frac{-\sin x(1-\sin x)-\cos x\cdot(-\cos x)}{(1-\sin x)^2}=\frac{-\sin x+\sin^2 x+\cos^2 x}{(1-\sin x)^2}=$$

$$=\frac{1-\sin x}{(1-\sin x)^2}=\frac{1}{1-\sin x}.$$

Solution of 9):

$$y'=\left(\sqrt{x}\,arctg\,x\right)'=\left(x^{\frac{1}{2}}\right)'arctg\,x+\sqrt{x}\cdot(arctg\,x)'=$$

$$=\frac{1}{2}x^{-\frac{1}{2}}\cdot arctg\,x+\sqrt{x}\cdot\frac{1}{1+x^2}=\frac{arctg\,x}{2\sqrt{x}}+\frac{\sqrt{x}}{1+x^2}.$$

Solution of 10):

$$y'=\left(x^2\log_3 x\right)'=\left(x^2\right)'\log_3 x+x^2\cdot\left(\log_3 x\right)'=$$

$$=2x\log_3 x+x^2\cdot\frac{1}{x}\log_3 e=2x\log_3 x+x\log_3 e=2x\log_3 x+\frac{x}{\ln 3}.$$

Answer: $1)(x-2)^2$; $2)1+\frac{1}{\sqrt{x}}$; $5)1-\cos x$;

$8)\,arc\sin x+\frac{x}{\sqrt{1-x^2}}$; $11)\,x^{n-1}(n\ln x+1)$;

$12)\,2^x\ln 2$; $13)\,10^x(1+x\ln 10)$;

$14)\,e^x(\cos x-\sin x)$.

Problem 355. Find the derivatives of the following complex functions:

$1)\,y=\sin 6x$; $2)\,y=6\cos\frac{x}{3}$; $3)\,y=\sin\sqrt{x}$;

$4)\,y=\frac{1}{\left(1-x^2\right)^5}$; $5)\,y=\sin^4 x$; $6)\,y=\frac{1}{(1+\cos 4x)^5}$;

$7)\,y=\ln\sin x$; $8)\,y=\ln^4\sin x$; $9)\,y=\ln ch x$;

$10)\,y=\log_3\left(x^2-1\right)$; $11)\,y=10^{2x-3}$;

$12)\,y=e^{arc\sin 2x}$; $13)\,y=\sin\left(e^{x^2+3x-2}\right)$;

$14)\,y=arc\cos\frac{2x-1}{\sqrt{3}}$; $15)\,y=arctg\left(x-\sqrt{1+x^2}\right)$.

Solution of 3): Use the formula $(\sin u)' = u' \cos u$:

$$y' = \left(\sin\sqrt{x}\right)' = \left(\sqrt{x}\right)' \cos\sqrt{x} = \frac{1}{2\sqrt{x}} \cdot \cos\sqrt{x}.$$

Solution of 6): At first apply the formula $\left((u)^n\right)' = nu^{n-1} \cdot u'$, then the formula $(\cos u)' = -u' \sin u$ and get:

$$y' = \left((1+\cos 4x)^{-5}\right)' = -5(1+\cos 4x)^{-6} \cdot (1+\cos 4x)' =$$

$$= -\frac{5}{(1+\cos 4x)^6} \cdot (-\sin 4x) \cdot (4x)' = \frac{20\sin 4x}{(1+\cos 4x)^6}.$$

Solution of 8): At first use the formula $\left(u^n\right)' = nu^{n-1} \cdot u'$, then $(\ln u)' = \frac{u'}{u}$ and get:

$$y' = \left(\ln^4 \sin x\right)' = 4\ln^3 \sin x \cdot (\ln \sin x)' = 4\ln^3 \sin x \cdot \frac{(\sin x)'}{\sin x} =$$

$$= 4\ln^3 \sin x \cdot \frac{\cos x}{\sin x} = 4ctg\, x \ln^3 \sin x.$$

Solution of 13): At first use the formula $(\sin u)' = u' \cos u$, then $\left(e^u\right)' = u'e^u$ and get:

$$y' = \left(\sin e^{x^2+3x-2}\right)' = \left(e^{x^2+3x-2}\right)' \cos e^{x^2+3x-2} =$$

$$= \left(x^2+3x-2\right)' e^{x^2+3x-2} \cdot \cos e^{x^2+3x-2} =$$

$$= (2x+3)e^{x^2+3x-2} \cdot \cos e^{x^2+3x-2}.$$

Answer: $1) 6\cos 6x$; $2) -2\sin\frac{x}{3}$; $4) \frac{10x}{\left(1-x^2\right)^6}$;

$5) 4\sin^3 x \cos x$; $7) ctg\, x$; $9) th\, x$;

$10) \frac{2x}{x^2-1}\log_3 e$; $11) 2 \cdot 10^{2x-3} \ln 10$;

$12)\frac{2}{\sqrt{1-4x^2}}e^{\arcsin 2x}$; $14)-\frac{2}{\sqrt{2+4x-4x^2}}$;

$15)\frac{1}{2\left(1+x^2\right)}$.

Problem 356. Find the value of the derivative of the function $y = f(x)$ at the point x_0:

$$1)\, y=\left(2-x^2\right)\cos x+2x\sin x,\ x_0=0;$$

$$2)\, y=arctg\, x\cdot arc\cos x,\ x_0=0;$$

$$3)\, y=\log_2 x\cdot\ln 2x,\ x_0=1;$$

$$4)\, y=x^5e^{-x},\ x_0=5.$$

Solution of 3): For finding the value of the function $f(x)$ at the point x_0 at first we must calculate its $f'(x)$ derivative and then in the expression of $f'(x)$ write x_0 instead of x.

$$y'(x)=\left(\log_2 x\cdot\ln 2x\right)'=\frac{1}{x}\log_2 e\cdot\ln 2x+\log_2 x\cdot\frac{2}{2x}=\frac{\ln 2x}{x\ln 2}+\frac{\log_2 x}{x}.$$

$$y'(x_0)=y'(1)=\frac{\ln 2\cdot 1}{1\cdot\ln 2}+\frac{\log_2 1}{1}=\frac{\ln 2}{\ln 2}+0=1.$$

Answer: $1)0$; $2)\frac{\pi}{2}$; $4)0$.

Problem 357. Find the right and left derivatives of the following functions at the indicated points:

$$1)\, f(x)=\left|2^x-2\right|,\ x=1;$$

$$2)\, f(x)=\left|x^2-5x+6\right|,\ x=2,\ x=3.$$

Solution of 1): At first let us find the right derivative of $f(x)$ at the point $x = 1$. For that, at the point $x = 1$ we can give to x a positive increment Δx.

$$f'_+(1)=\lim_{\substack{\Delta x\to 0\\ \Delta x>0}}\frac{f(1+\Delta x)-f(1)}{\Delta x}=\lim_{\substack{\Delta x\to 0\\ \Delta x>0}}\frac{\left|2^{1+\Delta x}-2\right|-\left|2^1-2\right|}{\Delta x}=$$

$$= \lim_{\substack{\Delta x\to 0\\ \Delta x>0}} \frac{\left|2^{1+\Delta x}-2\right|}{\Delta x} = \lim_{\Delta x\to 0}\frac{2^{1+\Delta x}-2}{\Delta x} = \lim_{\Delta x\to 0}\frac{2\cdot\left(2^{\Delta x}-1\right)}{\Delta x} = 2\ln 2 = \ln 4\,.$$

So, $f'_+(1)=\ln 4$.

If we give to x a negative increment Δx at the point $x = 1$, then as $2^{1+\Delta x} < 2$,

$$2^{1+\Delta x}-2<0 \Rightarrow \left|2^{1+\Delta x}-2\right| = -\left(2^{1+\Delta x}-2\right)\cdot$$

Therefore, we get

$$f'_-(1) = \lim_{\substack{\Delta x\to 0\\ \Delta x<0}} \frac{f(1+\Delta x)-f(1)}{\Delta x} = \lim_{\substack{\Delta x\to 0\\ \Delta x<0}} \frac{\left|2^{1+\Delta x}-2\right|-\left|2^{1}-2\right|}{\Delta x} =$$

$$= \lim_{\Delta x\to 0}\frac{-\left(2^{1+\Delta x}-2\right)}{\Delta x} = -2\lim_{\Delta x\to 0}\frac{2^{\Delta x}-1}{\Delta x} = -2\ln 2 = -\ln 4\,.$$

So, $f'_+(1)=\ln 4,\ \ f'_-(1)=-\ln 4$.

Answer: 2) $f'_-(2)=f'_-(3)=-1,\ \ f'_+(2)=f'_+(3)=1$.

Guideline: When calculating the left and right limits, for $2 < x < 3$ use $f(x) < 0$, when $x < 2$ or $x > 3$ use $f(x) > 0$.

Problem 358. Prove that *a)* the derivative of the even function is an odd function;

b) the derivative of the odd function is an even function.

Solution of *a)***:** When $f(x)$ is even, as $f(-x) = f(x)$ we can write:

$$f'(-x_0) = \lim_{\Delta x\to 0}\frac{f(-x_0+\Delta x)-f(-x_0)}{\Delta x} = \lim_{\Delta x\to 0}\frac{f(x_0-\Delta x)-f(x_0)}{\Delta x} =$$

$$= -\lim_{\Delta x\to 0}\frac{f(x_0-\Delta x)-f(x_0)}{-\Delta x} = -f'(x_0)\,.$$

As $f'(-x_0)=-f'(x_0)$ the function $f'(x)$ is an odd function.

Problem 359. Find $\dfrac{dy}{dx}$ for the following functions given in parametric form:

$$1)\, x=t^2,\ y=2t;\ 2)\, x=a\cos t,\ y=b\sin t;$$

$$3)\, x=\cos^3 t,\ y=\sin^3 t;\ 4)\, x=e^t\sin t,\ y=e^t\cos t.$$

Answer: 1) $\dfrac{dy}{dx}=\dfrac{1}{t}$; 2) $\dfrac{dy}{dx}=-\dfrac{b}{a}ctg\,t$; 3) $\dfrac{dy}{dx}=-tg\,t$;

4) $\dfrac{dy}{dx}=\dfrac{\cos t-\sin t}{\cos t+\sin t}$.

Guideline: Use (3.8).

Problem 360. Find the $y'(x)$ derivatives of the functions given in implicit form:

$$1)\frac{x^2}{a^2}+\frac{y^2}{b^2}=1;\ 2)\,x^4+y^4=x^2y^2;$$
$$3)\,2^x+2^y=2^{x+y};\ 4)\cos(xy)=x.$$

Solution of 3): differentiate the both hand sides of the equation $2^x+2^y=2^{x+y}$ with respect to x:

$$2^x\ln 2+\left(2^y\ln 2\right)y'=2^{x+y}\ln 2\cdot(1+y')\Rightarrow 2^x+2^y y'=2^{x+y}(1+y').$$

Solve the last equation with respect to y':

$$\left(2^{x+y}-2^y\right)y'=2^x-2^{x+y}\Rightarrow y'=\frac{2^x\left(1-2^y\right)}{2^y\left(2^x-1\right)}=2^{x-y}\cdot\frac{1-2^y}{2^x-1}.$$

So, $y'(x)=2^{x-y}\cdot\dfrac{1-2^y}{2^x-1}$.

Let us note that from the given equation we could find 2^y and 2^{-y} and write them in the last formula.

Answer: 1)$-\dfrac{b^2x}{a^2y}$; 2)$\dfrac{x}{y}\cdot\dfrac{y^2-2x^2}{2y^2-x^2}$; 4)$-\dfrac{1+y\sin(xy)}{x\sin(xy)}$.

Problem 361. Based on the theorem on differentiation of functions, show that the following formulas are true:

$$1)(\arcsin x)'=\frac{1}{\sqrt{1-x^2}};\ 2)(arc\,tg\,x)'=\frac{1}{1+x^2}.$$

Solution of 1): $y=\arcsin x$ is the inverse function of the function $x=\sin y$. By definition of differentiation of the odd function (see (3.5)) we can write:

$$(\arcsin x)' = \frac{1}{(\sin y)'} = \frac{1}{\cos y} = \frac{1}{\sqrt{1-\sin^2 y}} = \frac{1}{\sqrt{1-x^2}}.$$

Problem 362. Find the derivatives:

$$1)\, y = x^x,\ (x>0);\ 2)\, y = (\cos x)^{\sin x},\ \left(0 < x < \frac{\pi}{2}\right);$$

$$3)\, y = x^{e^x},\ (x>0).$$

Answer: $1)\left(x^x\right)' = x^x(1+\ln x);$

$$2)\left(\cos x^{\sin x}\right)' = \cos x^{\sin x}(\cos x \ln \cos x - tg\, x \cdot \sin x);$$

$$3)\, e^x \cdot x^{e^x} \cdot \left(\frac{1}{x} + \ln x\right).$$

Guideline: Use logarithmic derivative or formula (3.10).

Problem 363. Find the indicated order derivatives of the following functions:

$$1)\, y = \sin^2 x,\ y'' = ?\ 2)\, y = \sqrt{1+x^2},\ y'' = ?$$
$$3)\, y = xe^{-x},\ y^{(3)} = ?\ 4)\ y = x\sin x,\ y^{(3)} = ?$$

Answer: $1)\, 2\cos 2x; 2)\frac{1}{\left(1+x^2\right)^{3/2}}; 3)\, e^{-x}(3-x);$

$$4) -(x\cos x + 3\sin x).$$

Home tasks

Problem 364. By definition of derivative, find the derivatives of the functions:

$$1)\, y = e^{3x};\ 2)\, y = \sin(3x+1)\cdot$$

Answer: $1)\, 3e^{3x};\ 2)\, 3\cos(3x+1)$.

Problem 365. Find the derivatives of the following functions:

$$1)\ y=\frac{x^5}{5}-\frac{2x^3}{3}+x;\ 2)\ y=\frac{10}{x^3};$$

$$3)\ y=3x-6\sqrt{x};\ 4)\ y=x-tg\,x;$$

$$5)\ y=x\ln x;\ 6)\ y=x-arctg\,x;$$

$$7)\ y=x^2+2^x;\ 8)\ y=x^2\cdot 2^x;$$

$$9)\ y=x^3\log_5 x.$$

Answer: $1)\left(x^2-1\right)^2;2)-\dfrac{30}{x^4};3)3\left(1-\dfrac{1}{\sqrt{x}}\right);4)-tg^2x;$

$5)\ln x+1;6)\dfrac{x^2}{1+x^2};7)2x+2^x\ln 2;$

$8)\left(2x+x^2\ln 2\right)\cdot 2^x;9)x^2\log_5\left(ex^3\right).$

Problem 366. Find the derivatives:

$$1)\ y=\sin\frac{x}{2}+\cos\frac{x}{2};2)\ y=(1-5x)^4;3)\ y=\sqrt[3]{(4+3x)^2};$$

$$4)\ y=\sqrt{\cos 4x};5)\ y=ctg^3\frac{x}{3};6)\ y=\ln\left(e^{-x}+xe^{-x}\right);$$

$$7)\ y=arctg\frac{1+x}{1-x};8)\ y=\log_2^3(2x+3)^2;$$

$$9)\ y=e^{\sqrt{\ln\left(x^2+x+1\right)}};10)\ y=\ln tg\left(\frac{\pi}{4}+\frac{\sin x}{2}\right).$$

Answer: $1)\dfrac{1}{2}\left(\cos\dfrac{x}{2}-\sin\dfrac{x}{2}\right);2)-20(1-5x)^3;3)\dfrac{2}{\sqrt[3]{4+3x}};$

$4)-2tg4x\sqrt{\cos 4x};5)-\dfrac{ctg^2\dfrac{x}{3}}{\sin^2\dfrac{x}{3}};6)-\dfrac{x}{1+x};$

$7)\dfrac{1}{1+x^2};8)\dfrac{12}{\ln 2}\cdot\dfrac{\log_2^2(2x+3)^2}{2x+3};$

$9)\dfrac{(2x+1)e^{\sqrt{\ln\left(x^2+x+1\right)}}}{2\left(x^2+x+1\right)\sqrt{\ln\left(x^2+x+1\right)}};10)\dfrac{\cos x}{\cos(\sin x)}.$

Problem 367. Find the right and left derivatives of the function $f(x) = \ln|x|$ at the point $x = 1$.

Answer: $f'_+(1) = 1$, $f'_-(1) = -1$.

Problem 368. Prove that the derivative of a periodic function of period T is a periodic function of the same period.

Problem 369. Find $f'(a)$:

$$1)\ f(x) = \sin^2 x + \sin x^2, a = 0;$$

$$2)\ f(x) = \cos\frac{x}{5} + \cos\frac{5}{x}, a = 5.$$

Answer: 1) $f'(0) = 0$; 2) $f'(5) = 0$.

Problem 370. Find the first-order derivative of y with respect to x:

$$1)\ x = a(t - \sin t),\ y = a(1 - \cos t);$$

$$2)\ x = a\cos^3 t,\ y = b\sin^3 t;$$

$$3)\ x = e^t \cos t,\ y = e^t \sin t.$$

Answer: 1) $\frac{dy}{dx} = ctg\frac{t}{2}, (t \neq 2k\pi)$; 2) $\frac{dy}{dx} = -\frac{b}{a}tg\, t$;

3) $\frac{dy}{dx} = \frac{\cos t + \sin t}{\cos t - \sin t}$.

Problem 371. Find the derivatives of the following functions given in implicit form:

$$1)\frac{x^2}{a^2} - \frac{y^2}{b^2} = 1; 2)\ e^y + xy = e; 3)\ x^{\frac{2}{3}} + y^{\frac{2}{3}} = a^{\frac{2}{3}}.$$

Answer: 1) $y' = \frac{b^2 x}{a^2 y}$; 2) $y' = -\frac{y}{x + e^y}$; 3) $y' = -\sqrt[3]{\frac{y}{x}}$.

Problem 372. Based on definition of differentiation of inverse functions, prove that

$$1)\,(\arccos x)' = -\frac{1}{\sqrt{1-x^2}};\ 2)\,(arc\,ctg\,x)' = -\frac{1}{1+x^2};$$

$$3)\left(\log_a x\right)' = \frac{1}{x\ln a},\ (0 < a \neq 1).$$

Problem 373. Find the derivatives:

$$1)\,y = x^{2^x};\ 2)\,y = \left(\sqrt{x}\right)^{\sqrt[3]{x}};3)\,y = (\sin x)^{\arcsin x}.$$

Answer: $1)\,x^{2^x}\cdot 2^x\left(\frac{1}{x}+\ln x\cdot\ln 2\right);\ 2)\left(\sqrt{x}\right)^{\sqrt[3]{x}}\cdot\frac{3+\ln x}{6\sqrt[3]{x^2}};$

$$3)\,(\sin x)^{\arcsin x}\left(\frac{\ln\sin x}{\sqrt{1-x^2}}+\arcsin x\cdot ctg\,x\right).$$

Problem 374. Find the values of indicated order derivatives of the following functions at the given points:

$$1)\,y(x) = e^{2x}\cdot\sin 3x,\ y'(0), y''(0), y'''(0);$$
$$2)\,y(x) = \ln(x-1),\ y'''(2).$$

Answer: $1)\,y'(0) = 3, y''(0) = 12, y'''(0) = 9; 2)\,y'''(2) = 2.$

3.2 SOME APPLICATIONS OF DERIVATIVE AND DIFFERENTIAL

The value $f'(x_0)$ of the derivative of the function $y=f(x)$ differentiable at the point x_0 equals the angular coefficient of the tangent drawn at the point $M_0(x_0; f(x))$ to the curve of the graph of this function: $k=f'(x_0)$.

The equation of the straight line tangent to the graph of the function $y=f(x)$ at the point $M_0(x_0; f(x))$ is written in the form

$$y - f(x_0) = f'(x_0)(x - x_0)\,. \tag{3.11}$$

The straight line perpendicular to the tangent of the curve at the tangency point is said the normal of this curve at that point. The equation of the normal of the graph of the function $y=f(x)$ at the point $M_0(x_0; f(x))$ is

$$y - f(x_0) = -\frac{1}{f'(x_0)}\cdot(x - x_0)\,. \tag{3.12}$$

If the function $S(t)$ expresses the dependence of the S path of the moving point on time t, then $S'(t)$ indicates the rate of the moving point: $S'(t) = v$. This is the mechanical sense of derivative.

Sometimes, $f'(x_0)$ is considered as the rate of the function $f(x)$ at the point x.

Derivative has some other properties.

Now we give definition to the differential of a function.

It is known that for the function $y = f(x)$ differentiable on the interval (a, b) at any point x we can write

$$\Delta y = f'(x)\Delta x + \alpha(\Delta x)\cdot\Delta x,\ \lim_{\Delta x\to 0}\alpha(\Delta x) = 0 \tag{3.13}$$

Definition. The principal part of the increment of the function $y = f(x)$ at the point x linearly dependent on Δx is said a differential of $f(x)$ at the point x and is denoted by

$$d\,f(x) = f'(x)\Delta x \text{ or } d\,y = y'\Delta x. \tag{3.14}$$

When taking $f(x) = x$, as $f'(x) = 1 \Rightarrow df(x) = dx = 1\cdot\Delta x = \Delta x \Rightarrow \Rightarrow dx = \Delta x$ (4) is written as:

$$d\,f(x) = f'(x)\,d\,x. \tag{3.15}$$

Note that when x argument is dependent on other t variable, this differential is also in the form (3.15). This is called invariance property (invariability) of the form (3.15) of a differential.

The differential of function differential is said to be second-order or second differential of this function and is denoted by d^2y, $d^2f(x)$. The definition of higher order differentials is given in the same way. The n-th order differential of a function is found by the formula

$$d^n y = f^{(n)}(x)\,d\,x^n. \tag{3.16}$$

When the modulus of the argument increment $\Delta x = x - x_0$, is small, the value of Δy in small neighborhood of the point x_0 is close to dy: $\Delta y \approx dy$. Therefore, in (4), replacing $d\,y$ by $\Delta y = f(x) - f(x_0)$, we can write

$$\Delta y \approx f'(x_0)\Delta x \Rightarrow f(x) - f(x_0) \approx f'(x_0)\Delta x \Rightarrow$$

$$\Rightarrow f(x) \approx f(x_0) + f'(x_0)\Delta x. \tag{3.17}$$

When it becomes easy to find the values of the function $f(x)$ and its derivatives at the point x_0, for calculating approximate value of the function $f'(x)$ at the point x near to x_0, formula (3.17) is used.

Problems to be solved in auditorium

Problem 375. On which point of the parabola $y = -x^2 + 2x - 3$ does the tangent drawn to this parabola make an angle of 45° with positive direction of the abscissa axis?

Answer: (0.5; −2.25). ***Guideline:*** Find the x_0 abscissa of the tangent point from the relation $f'(x_0) = \text{tg}45°$.

Problem 376. Write to the graph of the function $y=f(x)$ the equation of the tangent and normal drawn at the point with the basis x_0:

$$1)\, y = x^3 + 2x^2 - 4x - 3,\ x_0 = -2;$$

$$2)\, y = tg2x,\ x_0 = 0; 3)\, y = e^{1-x^2},\ x_0 = -1.$$

Answer: $1)\, y-5=0, x+2=0;\ 2)\, y-2x=0,\ 2y+x=0;$

$3)\, 2x-y+3=0, x+2y-1=0.$

Problem 377. At the point $M_0\,(x_0; y_0)$ on the ellipse $\frac{x^2}{a^2}+\frac{y^2}{b^2}=1$ derive the equation of the straight line tangent to this ellipse.

Answer: $\frac{xx_0}{a^2}+\frac{yy_0}{b^2}=1$.

Guideline: Use $\frac{x_0^2}{a^2}+\frac{y_0^2}{b^2}=1$ and 1) in problem 360.

Problem 378. Write the equation of the tangent and normal drawn to the curve (cycloid) given by the equations $x = t \sin t$, $y = 1 - \cos t$ at the point corresponding to the value $t=\frac{\pi}{2}$ of the parameter of this curve.

Solution. For, $t=\frac{\pi}{2}$, $x\left(\frac{\pi}{2}\right)=\frac{\pi}{2}-\sin\frac{\pi}{2}=\frac{\pi}{2}-1,$

$$y\left(\frac{\pi}{2}\right)=1-\cos\frac{\pi}{2}=1\cdot$$

$$\frac{dy}{dx}=\frac{y_t'}{x_t'}=\frac{\sin t}{1-\cos t}\Rightarrow \left.\frac{dy}{dx}\right|_{t=\frac{\pi}{2}}=\frac{\sin\frac{\pi}{2}}{1-\cos\frac{\pi}{2}}=1.$$

Then the equation of the tangent is

$$y-1=1\cdot\left(x-\frac{\pi}{2}+1\right)\Rightarrow x-y-\frac{\pi}{2}+2=0\Rightarrow 2x-2y-\pi+4=0,$$

the equation of the normal is

$$y-1=-\frac{1}{1}\cdot\left(x-\frac{\pi}{2}+1\right)\Rightarrow y-1=-x+\frac{\pi}{2}-1=0\Rightarrow 2x+2y-\pi=0.$$

Problem 379. The rule of movement of a material point along a straight line is in the form $x=\frac{1}{4}t^4-4t^3+16t^2$.

1) At which moments of time the point is at the origin of coordinates?
2) At which moments of time, its motion direction coincides with the positive direction of OX axis?

Answer: 1) $t_1=0,\ t_2=8$; 2) $t\in(0;4)\cup(8;+\infty)$.

Guideline: 1) Solve the equation $x(t)=0$; 2) the inequality $dx/dt>0$.

Problem 380. Find differential of the following functions:

$$1)\ y=\frac{1}{2\sqrt{x}};\ 2)\ y=\sin x+\sqrt[3]{x};\ 3)\ y=\sqrt{\arcsin 2x}+a^{-x}.$$

Answer: $1)\ dy=-\frac{dx}{4x\sqrt{x}};\ 2)\ dy=\cos x dx+\frac{dx}{3\sqrt[3]{x^2}};$

$$3)\ dy=\frac{dx}{\sqrt{\arcsin 2x}\cdot\sqrt{1-4x^2}}-a^{-x}\ln a dx.$$

Problem 381. Find differentials at the given points:

$$1)\ d\left(\frac{1}{x}+\ln\frac{x-1}{x}\right),x=-1;\ 2)\ d\left(\frac{x^2 2^x}{x^x}\right),x=1,x=2.$$

Answer: $1)-\frac{1}{2}dx;\ 2)(2+\ln 4)dx,0.$

Problem 382. Calculate approximate values of the numbers:

$$1)\ \sqrt[4]{17}\ ;\ \ 2)\ arctg0.98\ ;\ \ 3)\ \sin 29^\circ.$$

Solution of 2): Take $f(x) = arctg\, x$, $x_0 = 1$, $x = 0.98$ and formula (7). As,

$$f(x_0) = arctg 1 = \frac{\pi}{4} \approx 0.785 \; ; \; f'(x) = \frac{1}{1+x^2} \Rightarrow$$

$$\Rightarrow f'(x_0) = \frac{1}{1+1} = \frac{1}{2}, \quad \Delta x = x - x_0 = 0.98 - 1 = -0.02,$$

$$\text{or } arctg\, 0.98 \approx 0.785 + \frac{1}{2}\cdot(-0.02) = 0.785 - 0.01 = 0.775.$$

So, arctg $0.98 \approx 0.775$.

Answer: 1) $\sqrt[4]{17} \approx 2.031$; 3) $\sin 29^\circ \approx 0.4848$.

Problem 383. Prove that at small values of Δx with respect to x_0 the following approximate calculation formula is valid

$$\sqrt[n]{x_0 + \Delta x} \approx \sqrt[n]{x_0} + \frac{\sqrt[n]{x_0}}{n x_0}\Delta x, \; (x_0 > 0)$$

Using this formula, calculate:

1) $\sqrt[3]{200}$; 2) $\sqrt[5]{243.45}$.

Answer: 1) $\sqrt[3]{200} \approx 5.8519$; 2) $\sqrt[5]{243.45} \approx 3.0011$.

Guideline: 1) $x_0 = 216$, $\Delta x = -16$; take

2) $x_0 = 243$, $\Delta x = 0.45$.

Problem 384. Find the indicated order differentials of the functions:

1) $y = (2x-1)^3, d^3y = ?$ 2) $y = e^{2x}, d^2y = ?$

3) $y = a\sin(bx+c), d^2y = ?$ 4) $y = \frac{\sin x}{x}, d^2y = ?$

Answer: 1) $48dx^3$; 2) $4e^{2x}dx^2$; 3) $-ab^2\sin(bx+c)dx^2$;

4) $\frac{(2-x^2)\sin x - 2x\cos x}{x^3}dx^2$.

Home tasks

Problem 385. Write the equation of the tangent and normal drawn to the graph of the function $y = f(x)$ at the point with x_0 abscissa:

$$1)\ y = x^2 - 5x + 4, x_0 = -1;\ 2)\ y = \sqrt{x}, x_0 = 4;$$
$$3)\ y = \ln x, x_0 = 1.$$

Answer: $1)\ 7x + y - 3 = 0,\ x - 7y + 71 = 0;$
$2)\ x - 4y + 4 = 0,\ 4x + y - 18 = 0;$
$3)\ x - y - 1 = 0,\ x + y - 1 = 0.$

Problem 386. Write the equations of a straight lines tangent to the curve given by the formula $x = t\cos t$, $y = t\sin t$, $t \in (-\infty, +\infty)$ at the origin of coordinates and at the point corresponding to the value $t = \frac{\pi}{4}$ of the parameter t.

Answer: $y = 0,\ (\pi + 4)x + (\pi - 4)y - \pi^2 \cdot \frac{\sqrt{2}}{4} = 0.$

Problem 387. Write the equations of the tangent and normal drawn to the curve $x^3 + y^2 + 2x - 6 = 0$ at the point with ordinate $y_0 = 3$.

Answer: $5x + 6y - 13 = 0$, $6x + 5y + 21 = 0$.

Problem 388. Derive the equation of the straight line tangent to this hyperbola at the point $M_0(x_0; y_0)$ on the hyperbola $\frac{x^2}{a^2} - \frac{y^2}{b^2} = 1$.

Answer: $\frac{xx_0}{a^2} - \frac{yy_0}{b^2} = 1$.

Problem 389. Two points act on the abscissa axis by the laws $x = 100 + 5t$ and $x = \frac{t^2}{2}$. When meeting each other, with what velocity do these points segregate from each other (time is measured with meter, time with seconds)?

Answer: $15\frac{m}{\text{sec}}$.

Guideline: Find meeting moment from the equation $100 + 5t = \frac{t^2}{2}$, then find velocities of each point at this moment.

Problem 390. Find the differentials of the functions:

$$1)\, y = e^{-x} + \ln x;\ 2)\, y = \arccos e^{x};\ 3)\, y = \sqrt{x} + 2\sqrt{x+\sqrt{x}}.$$

Answer: $1)\left(\frac{1}{x} - e^{-x}\right)dx;\ 2) -\frac{e^{x}}{\sqrt{1-e^{2x}}}dx;$

$$3)\frac{\sqrt{x+\sqrt{x}} + 2\sqrt{x} + 1}{2\sqrt{x\left(x+\sqrt{x}\right)}}dx.$$

Problem 391. Find the differentials of the function $y = arctg\frac{\ln x}{x}$ at the points $x = \frac{1}{e}$ and $x = e$.

Answer: $dy\Big|_{x=\frac{1}{e}} = \frac{2e^{2}}{e^{2}+1}dx, \qquad dy\Big|_{x=e} = 0.$

Problem 392. By replacing the differential by its increment find the approximate values of the following functions:

$$1)\ \arcsin 0.51;\quad 2)\ \sqrt[4]{15.8}.$$

Answer: 1) $0.508\left(\approx 29^{\circ}7'\right)$; 2) 1.9938.

Problem 393. Find second-order differentials of the functions:

$$1)\, y = 3^{-x^{2}};\ 2)\, y = \sqrt{1-x^{2}}\arcsin x;$$
$$3)\, y = \ln\left(x + \sqrt{1+x^{2}}\right).$$

Answer: $1)\, d^{2}y = 3^{-x^{2}}(\ln 9)\left(2x^{2}\ln 3 - 1\right)dx^{2};$

$$2)\, d^{2}y = -\frac{\sqrt{1-x^{2}}\,x + \arcsin x}{\left(1-x^{2}\right)^{\frac{3}{2}}}dx^{2};$$

$$3)\, d^{2}y = -x\left(1+x^{2}\right)^{-\frac{3}{2}}dx^{2}.$$

3.3 MAIN THEOREMS OF DIFFERENTIAL CALCULUS

Taylor's formula

The following Roll, Lagrange and Cauchy theorems are said to be main theorems of differential calculus or mean value theorems.

Theorem 1 (Roll's). If the function $y = f(x)$ is continuous on the interval $[a,b]$, differentiable on the interval (a,b) and satisfies the condition $f(a) = f(b)$, then there is at least one point ξ on the interval (a,b) such that $f'(\xi) = 0$.

Theorem 2 (of Lagrange). For the function $y = f(x)$ continuous on the interval $[a,b]$ and differentiable on the interval (a,b) there exists at least one point ξ on the interval (a,b) such that the equality

$$f(b) - f(a) = f'(\xi) \cdot (b - a) \tag{3.18}$$

is satisfied.

If we take $a = x$, $b = x + \Delta x$ we can show the point ξ between x and $x + \Delta x$ in the form $\xi = x + \theta\, \Delta x$, $(0 < \theta < 1)$. Then formula (1) may be written as follows:

$$f(x + \Delta x) - f(x) = \Delta x \cdot f'(x + \theta\, \Delta x) \tag{3.19}$$

Formulas (1), (2) are called Lagrange or finite increments formulas.

Theorem 3 (of Cauchy). If the functions $f(x)$ and $g(x)$ are continuous on the interval, are differentiable on the interval (a,b), and on this interval satisfy the condition $g(x) \neq 0$, then on the interval $[a,b]$ there exists at least one point ξ such that the formula

$$\frac{f(b) - f(a)}{g(b) - g(a)} = \frac{f'(\xi)}{g'(\xi)} \tag{3.20}$$

is true.

Theorem 4 (of Taylor). Assume that the function $y = f(x)$ has a derivative continuous to within $n + 1$ th order in certain vicinity of the point a (including $n + 1$). Then for every x contained in this vicinity of the point a the following formula is true

$$f(x) = f(a) + \frac{f'(a)}{1!}(x - a) + \frac{f^{(2)}(a)}{2!}(x - a)^2 + \ldots +$$

$$+\frac{f^{(n)}(a)}{n!}(x-a)^n + R_{n+1}(x)\,. \tag{3.21}$$

Here $R_{n+1}(x)$ is called a remainder term and this term in the Lagrange form is determined by the formula

$$R_{n+1}(x) = \frac{f^{(n+1)}(a+\theta(x-a))}{(n+1)!}(x-a)^{(n+1)},\ (0<\theta<1)\,. \tag{3.22}$$

(4) is said to be Taylor's formula for the function $f(x)$. The remainder term is written in Peano form as follows:

$$R_{n+1}(x) = 0\left((x-a)^n\right). \tag{3.23}$$

For $a=0$, the equality

$$f(x) = f(0) + \frac{f^1(0)}{1!}x + \frac{f^{(2)}(0)}{2!}x^2 + \ldots + \frac{f^{(n)}(0)}{n!}x^n + \frac{f^{(n+1)}(\theta x)}{(n+1)!}x^{n+1} \tag{3.24}$$

obtained from the Taylor formula is called Maclaurin formula for the function $f(x)$.

The Taylor formula is widely used in approximate calculations.

Problems to be solved in auditorium

Problem 394. Verify if the Roll theorem is true for the function $f(x) = x\left(x^2-1\right)$ on the intervals [–1;1] and [0;1].

Problem 395. Find such numbers on the intervals (–1;1) and (1;2) such that the tangents drawn to the graph of the function $f(x) = \left(x^2-1\right)(x-2)$ at the point with the abscissa equal to these numbers be parallel to the abscissa axis.

Answer: $\xi_1 = \frac{2-\sqrt{7}}{3},\ \xi_2 = \frac{2+\sqrt{7}}{3}$.

Guideline: For the function $f(x)$ verify the conditions of the Roll theorem on the interval [–1;1] and [1;2] and find the roots of the equation $f'(\xi) = 0$.

Problem 396. If $f(x) = x(x-1)(x-2)(x-3)$, prove that the three roots of the equation $f'(x) = 0$ are real numbers.

Guideline: On each interval [0;1], [1;2], [2;3] apply the Roll theorem to the function $f(x)$.

Problem 397. Apply the Lagrange theorem to the function $f(x)=\sqrt{3}x^3+3x$ on the interval [0;1] and find appropriate ξ point from the interval (0;1).

Answer: $\xi=\frac{1}{\sqrt{3}}$.

Problem 398. For the functions $f(x)=2x^3+5x+1$ and $g(x)=x^2+4$ write the Cauchy formula on the interval [0;2] and find the appropriate ξ point from the interval (0;2).

Answer: $\xi_1=\frac{1}{2},\ \xi_2=\frac{5}{3}$.

Problem 399. Decompose the following functions by the Maclaurin formula: 1) $y=e^x$; 2) $y=\cos x$; 3) $y=\ln(1+x)$.

Answer: 1) $e^x=1+\frac{x}{1!}+\frac{x^2}{2!}+\ldots+\frac{x^n}{n!}+\frac{e^{\theta x}}{(n+1)!}x^{n+1}$;

$$2)\cos x=1-\frac{x^2}{2!}+\frac{x^4}{4!}+\ldots+(-1)^n\frac{x^{2n}}{(2n)!}+\frac{x^{2n+2}}{(2n+2)!}\cos(\theta x+(n+1)\pi);$$

$$3)\ln(1+x)=x-\frac{x^2}{2}+\ldots+(-1)^{n-1}\frac{x^n}{n}+\frac{x^{n+1}}{(n+1)(1+\theta x)^{n+1}},\ (x>-1).$$

Problem 400. Decompose the following functions by the Maclaurin formula to $0(x^n)$-th order:

$$1)\ f(x)=e^{\frac{1}{2}x+2};\ 2)\ f(x)=\cos\left(2x+\frac{\pi}{4}\right);$$

$$3)\ f(x)=\ln(5-4x).$$

Solution of 3): In the decomposition of the formula $\ln(1+x)$

$\ln(1+x)=\sum_{k=1}^{n}\frac{(-1)^{k-1}x^k}{k!}+0(x^n)$ (see problem 399, 3) if we take x instead of x, then

$$\ln(1-x)=\sum_{k=1}^{n}\frac{(-1)^{k-1}\cdot(-1)^k x^k}{k}+0(x^n)=-\sum_{k=1}^{n}\frac{x^k}{k}+0(x^n).$$

Transform the given function in the form

$$\ln(5-4x)=\ln 5\left(1-\frac{4}{5}x\right)=\ln 5-\ln\left(1-\frac{4}{5}x\right)$$

and for obtaining decomposition of the function $\ln\left(1-\frac{4}{5}x\right)$ in the decomposition of ln(1 + x) instead of x we take $\frac{4}{5}x$:

$$\ln(5-4x)=\ln 5+\ln\left(1-\frac{4}{5}x\right)=\ln 5-\sum_{k=1}^{n}\frac{1}{k}\cdot\left(\frac{4}{5}\right)^k x^k+0(x^n).$$

Answer: 1) $e^{\frac{1}{2}x+2}=\sum_{k=0}^{n}\frac{e^2}{2^k k!}x^k+0(x^n)$;

2) $\cos\left(2x+\frac{\pi}{4}\right)=\sum_{k=0}^{n}\frac{2^k}{k!}\left[\cos\frac{\pi}{4}(2k+1)\right]\cdot x^k+0(x^n)$.

Problem 401. Using the Maclaurin formula, calculate the values of the following numbers to within 0.001:

$$1)\ \sqrt{e};\quad 2)\ \cos 10^\circ;\quad 3)\ \ln 1.05.$$

Solution of 1): For the function e^x in the Maclaurin formula (see problem 399 in 1) take $x=\frac{1}{2}$:

$$\sqrt{e}=1+\frac{1}{2!}+\frac{1}{2^2\cdot 2!}+\frac{1}{2^3\cdot 3!}+\ldots+\frac{1}{2^n\cdot n!}+R_{n+1},$$

$$\text{here } R_{n+1}=\frac{e^{\frac{\theta}{2}}}{2^{n+1}\cdot(n+1)!},\ (0<\theta<1).$$

$$0<\theta<1,\ 2<e<3\Rightarrow R_n<\frac{e^{\frac{1}{2}}}{2^{n+1}\cdot(n+1)!}<\frac{2}{2^{n+1}\cdot(n+1)!}\le\frac{1}{2^n\cdot(n+1)!}.$$

$$\text{So, } R_n<\frac{1}{2^n\cdot(n+1)!}.$$

For finding the value of $\sqrt{e}$ to within 0.001, we should choose n such that $R_n < 0.001$.

$$\text{for } n=3, \quad R_n < \frac{1}{2^3\cdot(3+1)!} = \frac{1}{8\cdot 24} = \frac{1}{192},$$

$$\text{for } n=4, \ R_n < \frac{1}{2^4\cdot(4+1)!} = \frac{1}{1920} < 0,001.$$

So, for finding approximate value of $\sqrt{e}$ to within required exactness, it suffices to take the first four terms in decomposition of $\sqrt{e}$ by the Maclaurin formula.

$$\sqrt{e} \approx 1+\frac{1}{2}+\frac{1}{2^2\cdot 2\ !}+\frac{1}{2^3\cdot 3\ !}+\frac{1}{2^4\cdot 4\ !} = 1.5000+1.1250+$$

$$+0.0208+0.0026 = 1.6484 \Rightarrow \sqrt{e} \approx 1.648.$$

Answer: 2) $\cos 10^\circ \approx 0.985$; 3) $\ln 1.05 \approx 0.049$.

Home tasks

Problem 402. At the edges of the interval $[-1;1]$ the function $f(x) = \dfrac{5-x^2}{x^4}$ takes equal values. Its derivative $f'(x)$ equals zero only at two $x = \pm\sqrt{10}$ points. But these points are not contained on the interval $[-1;1]$. Why the statement of the Roll theorem is not satisfied for this function?

Answer: The function $f(x)$ is discontinuous at the point $x \in [-1;1]$.

Problem 403. On the interval $(0;1)$ find a point ξ such that the tangent drawn to the graph of the function $y = x^3$ at the point $(\xi;\xi^3)$ be parallel to the intersection connecting the points $(0;0)$ and $(1;1)$.

Answer: $\xi = \dfrac{\sqrt{3}}{3}$.

Guideline: Use the Lagrange formula and its geometrical sense.

Problem 404. Applying the Lagrange theorem to the function $\dfrac{1}{x^\alpha}$ prove that for any $\alpha > 0$ and $n \in N$ the inequality

$$\frac{1}{n^{\alpha+1}} < \frac{1}{\alpha}\cdot\left(\frac{1}{(n-1)^\alpha} - \frac{1}{n^\alpha}\right)$$

is true.

Guideline: Write the Lagrange formula for the interval $n \in N$, $[n-1;n]$ and use

$$n-1<\xi<n \Rightarrow \frac{1}{n^{\alpha+1}}<\frac{1}{\xi^{\alpha+1}}.$$

Problem 405. Decompose the following function by the Maclaurin formula:

$$1)\, y=\sin x;\ 2)\, y=(1+x)^{\alpha},\ (x>-1).$$

Answer: $1) \sin x=\frac{x}{11}-\frac{x^3}{3!}+\frac{x^5}{5!}-\ldots+(-1)^n\frac{x^{2n+1}}{(2n+1)!}+$

$$+\frac{x^{2n+3}}{(2n+3)!}\sin\left(\theta x+(2n+3)\frac{\pi}{2}\right),\ 0<\theta<1;$$

$$2)(1+x)^{\alpha}=1+\frac{\alpha}{1!}x+\frac{\alpha(\alpha-1)}{2!}x^2+\ldots+\frac{\alpha(\alpha-1)\ldots(\alpha-n+1)}{n!}x^n+$$

$$+\frac{\alpha(\alpha-1)\ldots(\alpha-n)}{(n+1)!}(1+\theta x)^{\alpha-n-1}x^{n+1},\ (x>-1),\ 0<\theta<1.$$

Problem 406. Using the Maclaurin formula, calculate the approximate values of the following numbers to within 0.001:

$$1) \sin 1;\ 2) \sqrt[5]{33}.$$

Answer: 1) $\sin 1 \approx 0.842$; 2) $\sqrt[5]{33} \approx 2.012$.

3.4 OPENING OF UNCERTAINTIES. BERNOULLI-DE L'HOSPITAL RULE

When $\lim\limits_{x\to a} f(x)=\lim\limits_{x\to a} g(x)=0$ $\left(\lim\limits_{x\to a} f(x)=\lim\limits_{x\to a} g(x)=\infty\right)$, the ratio $\frac{f(x)}{g(x)}$ is said to be uncertainty in the form $0/0\ (\infty/\infty)$. In this case the calculation of the limit $\lim\limits_{x\to a}\frac{f(x)}{g(x)}$ is said to be opening of uncertainty of in the form $0/0\ (\infty/\infty)$.

One of the methods for opening uncertainties of the form $\frac{0}{0}$ and $\frac{\infty}{\infty}$ is the Bernoulli-de L'Hospital rule expressed by the equality

$$\lim_{x\to a}\frac{f(x)}{g(x)}=\lim_{x\to a}\frac{f'(x)}{g'(x)},\quad (g'(x)\neq 0). \tag{3.25}$$

The validity of the equality $\lim_{x\to 0}\frac{\sin x}{x}=1$ may be easily shown by the Bernoulli-de L'Hospital rule:

$$\lim_{x\to 0}\frac{\sin x}{x}=\lim_{x\to 0}\frac{(\sin x)'}{(x)'}=\lim_{x\to 0}\frac{\cos x}{1}=\frac{1}{1}=1.$$

In some cases, Bernoulli-de L'Hospital rule should be successively applied several times. For example, when calculating the limit

$$\lim_{x\to 0}\frac{x-\sin x}{x^3}=\lim_{x\to 0}\frac{(x-\sin x)'}{(x^3)'}=\lim_{x\to 0}\frac{1-\cos x}{3x^2}=\lim_{x\to 0}\frac{(1-\cos x)'}{(3x^2)'}=$$

$$=\lim_{x\to 0}\frac{\sin x}{6x}=\frac{1}{6}\lim_{x\to 0}\frac{\sin x}{x}=\frac{1}{6}\cdot 1=\frac{1}{6}$$

the Bernoulli rule was used twice.

For

$$\lim_{x\to a}f(x)=0,\ \lim_{x\to a}g(x)=\infty,$$

using the relation

$$\lim_{x\to a}f(x)\cdot g(x)=\lim_{x\to a}\frac{f(x)}{\frac{1}{g(x)}}\quad\text{or}\quad \lim_{x\to a}f(x)\cdot g(x)=\lim_{x\to a}\frac{g(x)}{\frac{1}{f(x)}}$$

the uncertainty of the form $0\cdot\infty$ is reduced to uncertainty of the form $\frac{0}{0}$ or $\frac{\infty}{\infty}$

For $\lim_{x\to a}f(x)=\lim_{x\to a}g(x)=\infty$ for opening the uncertainty of the form $\infty-\infty$ at first we must write the difference $f(x)-g(x)$ in the form $f(x)-g(x)=f(x)\left(1-\frac{g(x)}{f(x)}\right)$. If $\lim_{x\to a}\frac{g(x)}{f(x)}\neq 1$, then $\lim_{x\to a}(f(x)-g(x))=\infty$ should be taken. For $\lim_{x\to a}\frac{g(x)}{f(x)}=1$, the uncertainty of the form $\infty-\infty$ is reduced to uncertainty of the form $\infty\cdot 0$.

The uncertainties of the form 0°, ∞°, 1^∞ encountered when calculating the limits of the form $\lim\limits_{x\to a} u(x)^{v(x)}$, together with the relation ln $y = v(x)$ ln $u(x)$ obtained from the equality $y = u(x)^{v(x)}$ is reduced to uncertainties of the form ∞ · 0. After calculating $\lim\limits_{x\to a} \ln y$, the limit $\lim\limits_{x\to a} y = \lim\limits_{x\to a} u(x)^{v(x)}$ is found.

Problems to be solved in auditorium

Problem 407. Open the uncertainties of the form $0/0$:

$$1)\lim_{x\to 1}\frac{3x^2+4x-7}{2x^2+3x-5};\ 2)\lim_{x\to 0}\frac{\ln\cos x}{\ln\cos 3x};$$

$$3)\lim_{x\to 0}\frac{\ln(1+x)-x}{tg^2 x};\ 4)\lim_{x\to 0}\frac{\ln\cos ax}{x^2};$$

$$5)\lim_{x\to 1}\frac{x^5-3x^2+7x-5}{x^4-5x+4}.$$

Solution of 3):

$$\lim_{x\to 0}\frac{\ln(1+x)-x}{tg^2 x}=\lim_{x\to 0}\frac{(\ln(1+x)-x)'}{(tg^2 x)'}=\lim_{x\to 0}\frac{\frac{1}{1+x}-1}{2tg\,x\cdot\frac{1}{\cos^2 x}}=$$

$$=\frac{1}{2}\lim_{x\to 0}\frac{-\frac{x}{1+x}}{\frac{\sin x}{\cos^3 x}}=\frac{1}{2}\lim_{x\to 0}\frac{\cos^3 x}{1+x}\cdot\frac{x}{\sin x}=-\frac{1}{2}\cdot 1\cdot 1=-\frac{1}{2}.$$

Answer: $1)\frac{10}{7};\ 2)\frac{1}{9};\ 4)-\frac{a^2}{2};\ 5)-6.$

Problem 408. Open the uncertainties of the form ∞/∞:

$$1)\lim_{x\to +0}\frac{\ln x}{\ln\sin x};\ 2)\lim_{x\to\frac{\pi}{2}+0}\frac{\ln\left(x-\frac{\pi}{2}\right)}{tg\,x};\ 3)\lim_{x\to -\infty}\frac{e^x}{x^3};$$

$$4)\lim_{x\to\infty}\frac{\ln x}{x};\ 5)\lim_{x\to 0}\frac{\ln x}{ctg\,x};\ 6)\lim_{x\to\frac{\pi}{2}}\frac{tg\,x}{tg\,3x}.$$

Solution 6):

$$\lim_{x\to\frac{\pi}{2}}\frac{tg\,x}{tg\,3x}=\lim_{x\to\frac{\pi}{2}}\frac{(tg\,x)'}{(tg\,3x)'}=\lim_{x\to\frac{\pi}{2}}\frac{\frac{1}{\cos^2 x}}{\frac{3}{\cos^2 3x}}=\frac{1}{3}\lim_{x\to\frac{\pi}{2}}\frac{\cos^2 3x}{\cos^2 x}=$$

$$=\frac{1}{3}\lim_{x\to\frac{\pi}{2}}\frac{\frac{1}{2}(1+\cos 6x)}{\frac{1}{2}(1+\cos 2x)}=\frac{1}{3}\lim_{x\to\frac{\pi}{2}}\frac{(1+\cos 6x)'}{(1+\cos 2x)'}=\frac{1}{3}\lim_{x\to\frac{\pi}{2}}\frac{-6\sin 6x}{-2\sin 2x}=$$

$$=\lim_{x\to\frac{\pi}{2}}\frac{\sin 6x}{\sin 2x}=\lim_{x\to\frac{\pi}{2}}\frac{(\sin 6x)'}{(\sin 2x)'}=\lim_{x\to\frac{\pi}{2}}\frac{6\cos 6x}{2\cos 2x}=\frac{3\cos 3\pi}{\cos\pi}=3.$$

Answer: 1) 1; 2) 0; 3) 0; 4) 0; 5) 0.

Problem 409. Open the uncertainties of the form $\infty\cdot 0$:

$$1)\lim_{x\to\infty}x\left(e^{\frac{1}{x}}-1\right);\ 2)\lim_{x\to+0}x\ln^3 x;$$

$$3)\lim_{x\to 0}\left(e^x+e^{-x}-2\right)ctg\,x;4)\lim_{x\to\infty}x\sin\frac{a}{x}.$$

Solution of 4):

$$\lim_{x\to\infty}x\sin\frac{a}{x}=\lim_{x\to\infty}\frac{\sin\frac{a}{x}}{\frac{1}{x}}=\lim_{x\to\infty}a\cdot\frac{\sin\frac{a}{x}}{\frac{a}{x}}=a\lim_{y\to 0}\frac{\sin y}{y}=a\cdot 1=a.$$

Answer: 1) 1; 2) 0; 3) 0.

Problem 410. Open the uncertainties of the form $\infty-\infty$:

$$1)\lim_{x\to 0}\left(\frac{1}{\sin x}-\frac{1}{x}\right);\ 2)\lim_{x\to 0}\left(\frac{1}{x}-\frac{1}{e^x-1}\right);\ 3)\lim_{x\to 0}\left(\frac{1}{x\,arctg\,x}-\frac{1}{x^2}\right);$$

$$4)\lim_{x\to 1}\left(\frac{1}{2\left(1-\sqrt{x}\right)}-\frac{1}{3\left(1-\sqrt[3]{x}\right)}\right);\ 5)\lim_{x\to\frac{\pi}{2}}\left(\frac{x}{ctg\,x}-\frac{\pi}{2\cos x}\right).$$

Solution of 5): As $\lim\limits_{x\to\frac{\pi}{2}}\frac{x}{ctg\,x}=\infty,\ \lim\limits_{x\to\frac{\pi}{2}}\frac{\pi}{2\cos x}=\infty$, the given limit is an uncertainty of the form $\infty-\infty$.

Using the notation $\frac{x}{ctg\,x}-\frac{\pi}{2\cos x}=\frac{2x-\pi\cdot\frac{1}{\sin x}}{2ctg\,x}$ transform it into the uncertainty of the form $\frac{0}{0}$ and apply to the obtained Bernoulli-de L'Hospital rule:

$$\lim_{x\to\frac{\pi}{2}}\left(\frac{x}{ctg\,x}-\frac{\pi}{2\cos x}\right)=\lim_{x\to\frac{\pi}{2}}\left(\frac{2x-\pi\cdot\frac{1}{\sin x}}{2ctg\,x}\right)=\lim_{x\to\frac{\pi}{2}}\frac{\left(2x-\pi\cdot\frac{1}{\sin x}\right)'}{(2ctg\,x)'}=$$

$$=\lim_{x\to\frac{\pi}{2}}\frac{2+\pi\cdot\frac{\cos x}{\sin^2 x}}{-2\cdot\frac{1}{\sin^2 x}}=\lim_{x\to\frac{\pi}{2}}\frac{2\sin^2 x+\pi\cos x}{-2}=-1.$$

Answer: 1) 0; 2) $\frac{1}{2}$; 3) $\frac{1}{3}$; 4) $\frac{1}{12}$.

Problem 411. Calculate the limits that relate to one of the uncertainties 0°, ∞°, 1^{∞}:

$$1)\ \lim_{x\to+0}x^{\sin x};\ 2)\ \lim_{x\to+\infty}x^{\frac{1}{x}};\ 3)\lim_{x\to1}x^{\frac{1}{1-x}};\ 4)\lim_{x\to0}\left(\frac{\sin x}{x}\right)^{\frac{1}{x^2}};$$

$$5)\ \lim_{x\to0}\left(\frac{\pi}{2}\arccos x\right)^{\frac{1}{x}};\ 6)\ \lim_{x\to+0}(1+x)^{\ln x}.$$

Solution of 4): As $\lim\limits_{x\to0}\frac{\sin x}{x}=1,\ \lim\limits_{x\to0}\frac{1}{x^2}=\infty$, the given limit is an uncertainty of the form 1^{∞}. Designate $y=\left(\frac{\sin x}{x}\right)^{\frac{1}{x^2}}$. Then

$$\ln y=\frac{1}{x^2}\ln\frac{\sin x}{x}\Rightarrow\lim_{x\to0}\ln y=\lim_{x\to0}\frac{1}{x^2}\cdot\ln\frac{\sin x}{x}=\lim_{x\to0}\frac{\ln\frac{\sin x}{x}}{x^2}=$$

$$= \lim_{x\to 0}\frac{\left(\ln\dfrac{\sin x}{x}\right)'}{\left(x^2\right)'} = \lim_{x\to 0}\frac{\left(\dfrac{\sin x}{x}\right)'\Big/\left(\dfrac{\sin x}{x}\right)}{2x} = \lim_{x\to 0}\frac{x\cos x-\sin x}{2x^2\sin x} =$$

$$= \lim_{x\to 0}\frac{\left(x\cos x-\sin x\right)'}{\left(2x^2\sin x\right)'} = \lim_{x\to 0}\frac{\cos x - x\sin x-\cos x}{4x\sin x+2x^2\cos x} =$$

$$= -\frac{1}{2}\lim_{x\to 0}\frac{\sin x}{2\sin x+x\cos x} = -\frac{1}{2}\lim_{x\to 0}\frac{\left(\sin x\right)'}{\left(2\sin x+x\cos x\right)'} =$$

$$= -\frac{1}{2}\lim_{x\to 0}\frac{\cos x}{2\cos x+\cos x-x\sin x} = -\frac{1}{2}\lim_{x\to 0}\frac{\cos x}{3\cos x-x\sin x} = -\frac{1}{2}\cdot\frac{1}{3} = -\frac{1}{6}.$$

So, we get $\lim\limits_{x\to 0}\ln y = -\dfrac{1}{6} \Rightarrow \lim\limits_{x\to 0} y = e^{-\frac{1}{6}}$. As $y=\left(\dfrac{\sin x}{x}\right)^{\frac{1}{x^2}}$ we find $\lim\limits_{x\to 0}\left(\dfrac{\sin x}{x}\right)^{\frac{1}{x^2}} = e^{-\frac{1}{6}} = \dfrac{1}{\sqrt[6]{e}}$.

Answer: $1)1;\ 2)1;\ 3)\dfrac{1}{e};\ 5)e^{-\frac{\pi}{2}};\ 6)1$.

Problem 412. Using decomposition by the Maclaurin formula, calculate the following limits:

$$1)\lim_{x\to 0}\frac{\ln(1+x)-x}{x^2};\ 2)\lim_{x\to 0}\frac{e^x-1-x}{x^2};\ 3)\lim_{x\to 0}\frac{\cos x-1+\dfrac{x^2}{2}}{x^4};$$
$$4)\lim_{x\to 0}\frac{1-\cos x}{x^2+x^3};\ 5)\ \lim_{x\to 0}\frac{tg\,x-\sin x}{x^3};\ 6)\lim_{x\to 1}\frac{x-1-\sin(2x-2)}{x-1+\sin(3x-3)}.$$

Solution of 6): The given limit is an uncertainty of the form $0/0$. We can calculate it by the Bernoulli-de L'Hospital rule. Here this limit will be calculated by using decomposition of the sin x function in Maclaurin's formula. We can write

$$\sin x = \frac{x}{1!}-\frac{x^3}{3!}+\frac{x^5}{5!}+\ldots+(-1)^n\frac{x^{2n+1}}{(2n+1)!}+0\left(x^{2n+2}\right)\Rightarrow$$

$$\Rightarrow \sin x = x + 0(x),\ \ x \to 0 \ \text{ for}$$

$$\sin(2x-2) = \sin 2(x-1) = 2(x-1) + 0(|x-1|),\ x \to 1;$$

$$\sin(3x-3) = \sin 3(x-1) = 3(x-1) + 0(|x-1|),\ x \to 1.$$

Take this expansion into account in the given limit:

$$\lim_{x\to 1}\frac{x-1-\sin(2x-2)}{x-1+\sin(3x-3)} = \lim_{x\to 1}\frac{x-1-2(x-1)-0(|x-1|)}{x-1+3(x-1)+0(|x-1|)} =$$

$$= \lim_{x\to 1}\frac{-(x-1)-0(|x-1|)}{4(x-1)+0(|x-1|)} = \lim_{x\to 1}\frac{-1-\dfrac{0(|x-1|)}{x-1}}{4+\dfrac{0(|x-1|)}{x-1}} = \frac{-1+0}{4+0} = -\frac{1}{4}.$$

Answer: 1)$-\frac{1}{2}$; 2)$\frac{1}{2}$; 3)$\frac{1}{24}$; 4)$\frac{1}{2}$; 5)$\frac{1}{2}$.

Guideline: Decomposition of the function $tg\,x$ in Maclaurin formula is of the form $tg\,x = x + \frac{x^3}{3} + \frac{2}{15}x^5 + 0(x^6), x \to 0$.

Home tasks

Problem 413. Open the uncertainties of the form $0/0$ or ∞/∞:

$$1)\lim_{x\to 1}\frac{x^5-1}{2x^3-x-1};\ 2)\lim_{x\to 0}\frac{x-arctg\,x}{x^3};\ 3)\lim_{x\to\infty}\frac{\ln x}{\sqrt{x}};$$

$$4)\lim_{x\to 0}\frac{\sin x - x\cos x}{x^3};\ 5)\lim_{x\to 1}\frac{x^{10}-10x+9}{x^5-5x+4};\ 6)\lim_{x\to 0}\frac{e^{2x}-1}{\arcsin 3x};$$

$$7)\lim_{x\to 0}\frac{\ln\cos ax}{\ln\cos bx},(b\neq 0);\ 8)\lim_{x\to\frac{\pi}{4}}\frac{ctg\,x-1}{\sin 4x};\ 9)\lim_{x\to\infty}\frac{\ln x}{x^m},(m>0).$$

Answer: 1)1; 2)$\frac{1}{3}$; 3)0; 4)$\frac{1}{3}$; 5)4,5; 6)$\frac{2}{3}$; 7)$\frac{a^2}{b^2}$; 8)$\frac{1}{2}$; 9)0.

Problem 414. Open the uncertainties of the form $0 \cdot \infty$:

$$1)\lim_{x\to 0} x\ln x;\ 2)\lim_{x\to\pi}(\pi - x)tg\frac{x}{2};3)\lim_{x\to 0}\sin x\cdot\ln ctg\,x;\ 4)\lim x^n\cdot e^{-x^2}.$$

Answer: 1)0; 2)2; 3)0; 4)0.

Problem 415. Open the uncertainties of the form $\infty - \infty$:

$$1)\lim_{x\to\infty}\left(\sqrt{x^2+3x}-x\right);\ 2)\lim_{x\to 1}\left(\frac{1}{x-1}-\frac{2}{x^2-1}\right);$$

$$3)\lim_{x\to\infty}\left(\sqrt{x^2+x+1}-\sqrt{x^2-x}\right);\ 4)\lim_{x\to 2}\left(\frac{1}{x-2}-\frac{12}{x^3-8}\right);$$

$$5)\lim_{x\to 0}\left(\frac{1}{\sin^2 x}-\frac{1}{4\sin^2\frac{x}{2}}\right);\ 6)\lim_{x\to 0}\left(\frac{1}{arctg\,x}-\frac{1}{x}\right)$$

$$7)\lim_{x\to 0}\left(\frac{1}{x^2}-ctg^2 x\right).$$

Answer: 1)1,5; 2)$\frac{1}{2}$; 3)1; 4)$\frac{1}{2}$; 5)$\frac{1}{4}$; 6)0; 7)$\frac{2}{3}$.

Problem 416. Find the limits related to one of the uncertainties 0°, ∞°, 1^∞:

$$1)\lim_{x\to+0}(\arcsin x)^{tgx};\ 2)\lim_{x\to\frac{\pi}{2}-0}(\pi-2x)^{\cos x};\ 3)\lim_{x\to+0}\left(x+2^x\right)^{\frac{1}{x}};$$

$$4)\lim_{x\to+0}(ctg\,x)^{\frac{1}{\ln x}};\ 5)\lim_{x\to\frac{\pi}{2}-0}(tg\,x)^{2x-\pi};\ 6)\lim_{x\to 0}\left(e^x+x\right)^{\frac{1}{x}}.$$

Answer: 1)1; 2)1; 3)2; 4)$\frac{1}{e}$; 5)1; 6)e^2.

Problem 417. Using expansion in Maclaurin formula, calculate the following limits:

$$1)\lim_{x\to 0}\frac{tg\,x-x}{\sin x-x};\ 2)\lim_{x\to 0}\frac{arctg\,x-\arcsin x}{tg\,x-\sin x}.$$

Answer:1) –2.

Guideline: Take $\sin x = x-\frac{x^3}{6}+o\left(x^4\right)$, $tg\,x = x+\frac{x^3}{3}+o\left(x^4\right)$;

2)–1.

Guideline: Take $\arcsin x = x - \frac{x^3}{3} + o\left(x^4\right)$,

$$arctg\, x = x + \frac{x^3}{6} + o\left(x^4\right).$$

KEYWORDS

- **increment**
- **derivative**
- **differential**
- **tangent**
- **indefinitenesses**

CHAPTER 4

STUDYING FUNCTIONS OF DIFFERENTIAL CALCULUS AND THEIR APPLICATION TO CONSTRUCTION OF GRAPHS

CONTENTS

ABSTRACT

In this chapter, we give brief materials on finding the monotonicity intervals, extremum, the least the greatness values on the segment of a function, the direction of convexity of functions graph, turning point of the graph, asymptotes of functions graph, and construction of function's graph based on the researches, and 21 problems.

4.1 FINDING THE MONOTONICITY INTERVALS, EXTREMUM, THE LARGEST AND LEAST VALUES OF FUNCTION

Finding of intervals where a differentiable function is monotone, is based on two theorems.

Theorem 1. Necessary and sufficient condition for the function $y = f(x)$ differentiable on the interval (a, b) to be non-decreasing (non-increasing) on this interval, is $f'(x) \geq 0$ ($f'(x) \leq 0$) at all points of the interval (a, b).

Theorem 2. Differentiable function whose derivative is positive (negative) at all points of interval (a, b), is increasing (decreasing) on this interval.

Thus, monotonicity intervals of differentiable function will be the intervals where the first-order derivative of this function does not change its sign. For example, when as for the function $f(x) = x^3 - 3x^2 - 4$ when $f'(x) = 3x^2 - 6x = 3x\ (x - 2)$, $x \in (-\infty; 0) \cup (2; +\infty)$, $f'(x) > 0$, when $x \in (0;2)$, $f'(x) < 0$, this function increases on the interval $(-\infty; 0) \cup (2; +\infty)$; decreases on the interval $(0;2)$.

Now let us be acquainted with the notion of extremum of the function.

Assume that the function $f(x)$ was determined in certain neighborhood of the point c.

Definition. If the point c has a neighborhood such that for all x points from this neighborhood ($f(x) \leq f(c)$) $f(x) \geq f(c)$, then for $f(x)$ it is said that at the point c the function $f(x)$ has a local maximum (minimum). Local maximum and local minimum are called local extremum.

The following statement is true.

Theorem 3. (Necessary condition for the existence of local extremum). If the function $f(x)$ differentiable at the point c has extremum at this point, then $f'(c) = 0$.

This theorem is called Fermat's theorem.

$f'(c) = 0$ is necessary but not sufficient for $f(x)$ to have extremum at the point c. For example, for the function $f(x) = x^3$ as $f'(x) = 3x^2 \Rightarrow f'(0) = 0$ this function has not extremum at the point $x = 0$.

The points with $f'(x) = 0$ are called stationary points of the function $f(x)$. The points where the first-order derivative of the continuous function vanishes or when it has no first-order derivative, are called critical points of this function.

Existence of extremum at the critical point is determined by one of the following theorems.

Theorem 4. (The first sufficient condition for the existence of extremum). Assume that the function $y = f(x)$ is continuous in the neighborhood of the critical point c and differentiable at the same neighborhood (except the point c). If at the points from this neighborhood satisfying the condition $x < c, f'(x) > 0$, $(f'(x) < 0)$ at the point satisfying the condition $x > c$, $f'(x) < 0$ $(f'(x) > 0)$, then the function $f(x)$ has local maximum (minimum) at the point c. If in the neighborhood of the point c from the left or right of c the function $f'(x)$ has the same sign, then the function $f(x)$ has no extremum at the point c.

Theorem 5. (The second sufficient condition for the existence of extremum). If at the stationary point c the function $y = f(x)$ has the second-order derivative, for $f^{(2)}(c) < 0$ at the point c the function has local maximum, for $f^{(2)}(c) > 0$ at the point c the function has local minimum.

Theorem 6. Assume that at the point c the function $y = f(x)$ has continuous derivatives to within nth order (including n) and derivatives satisfying the conditions $f'(c) = f^{(2)}(c) = \ldots = f^{(n-1)}(c) = 0$, $f^{(n)}(c) \neq 0$. Then, if $f^{(n)}(c) < 0$ and n is an even number, at the point c it has local minimum. When $f^{(n)}(c) > 0$ and n is an odd number, this point has no local extremum.

At last, let us show the rule for finding the least and greatest values on this interval of the function continuous on an interval.

For finding the largest (least) values of the function $f(x)$ continuous on the interval $[a, b]$, we should find the values $f(a), f(b)$ of the function $f(x)$ at edge points of this interval and the values of $f(x)$ at the critical points contained on the interval $[a, b]$ and take the largest (least) of the found values.

Problems to be solved in auditorium

Problem 418. Find monotonicity intervals of the functions:

$$1)\, f(x)=4x^3-21x^2+18x+7;\ 2)\, f(x)=x^2e^{-x^2};$$
$$3)\, f(x)=2x^2-\ln x;\ 4)\, f(x)=\sqrt{8x^2-x^4}.$$

Solution of 4): At first find the domain of definition of the function:

$$8x^2-x^4\geq 0\Rightarrow x^2\left(8-x^2\right)\geq 0\Rightarrow -2\sqrt{2}\leq x\leq 2\sqrt{2}.$$

Calculate $f'(x)$:

$$f'(x)=\frac{16x-4x^3}{2\sqrt{8x^2-x^4}}=\frac{2x(2+x)(2-x)}{\sqrt{8x^2-x^4}}.$$

By the intervals method, we can determine that for $-2\sqrt{2}<x<-2$ or $0<x<2, f'(x)>0$, for $-2<x<0$ or $2<x<2\sqrt{2}$ $f'(x)<0$. By Theorem 2, the function $f(x)$ increases on the interval $\left(-2\sqrt{2};-2\right)$ and (0;2), decreases on the intervals (–0;2) and $\left(2;2\sqrt{2}\right)$.

Answer: 1) $\left(-\infty;\frac{1}{2}\right)$ and (3; + ∞) is an increase, $\left(\frac{1}{2};3\right)$ is a decrease; 2) (–∞; –1) and (0;1) is an increase, (–1;0) and (1; + ∞) is a decrease; 3) $\left(\frac{1}{2};+\infty\right)$ is an increase, $\left(0;\frac{1}{2}\right)$ is a decrease interval.

Problem 419. Find the monotonicity intervals of the function $y(x)$:

$$1)\, x^2y^2+y^2=1, y>0;\ 2)\, x=\frac{e^{-t}}{1-t}, y=\frac{e^t}{1-t}, t>1;$$

$$3)\, y=f(x)=\begin{cases}\frac{1}{e},\ for\ x<0,\\ \frac{\ln x}{x},\ for\ x\geq e.\end{cases}$$

Solution of 3): The given function is differentiable on the real axis and

$$f'(x) = \begin{cases} 0, & for \ \ x < e, \\ \dfrac{1 - \ln x}{x^2}, & for \ \ x \geq e. \end{cases}$$

It is seen from the expression of $f'(x)$ that at any values of x, $f'(x) \leq 0$. By Theorem 1 the given function is non-increasing on the real axis. More exactly, the function $f(x)$ is constant on the interval $(-\infty; e)$, is decreasing on the interval $(e; +\infty;)$.

Answer: 1) $(-\infty; 0)$ is an increase, $(0; +\infty;)$ is a decrease; 2) $(-\infty; -e^{-2})$ is an increase, $(-e^{-2}; 0)$ is a decrease interval.

Guideline: Using the expression of $x(t)$ and the relations $\lim\limits_{t \to 1+0} x(t) = -\infty$, $\lim\limits_{t \to \infty} x(t) = 0$ express the solutions of the inequalities $y'_x > 0$, $y'_x < 0$ expressed by the variable t, by the variable x.

Problem 420. Study the extremum of the function $y(x)$:

$$1)\ y = x\sqrt{1 - x^2};\ 2)\ y = \frac{x}{\ln x};\ 3)\ y = x - 2\sin x;$$

$$4)\ y = \ln x - arctg\, x;\ 5)\ y = x^{\frac{2}{3}};\ 6)\ y = |x - 3|;$$

$$7)\ y = (x-3)^2 e^{|x|};\ 8)\ y = \begin{cases} \dfrac{x}{1 + e^{\frac{1}{x}}}, x \neq 0 \\ 0, \quad x = 0. \end{cases}$$

Solution 7): The given function is everywhere differentiable except the point $x = 0$ and for the points $x \neq 0$, we can write:

$$y' = \begin{cases} 2(x-3)e^{-x} - (x-3)^2 e^{-x} = (x-3)(5-x)e^{-x}, & for \ \ x < 0 \\ 2(x-3)e^{x} - (x-3)^2 e^{x} = (x-3)(x-1)e^{x}, & for \ \ x > 0. \end{cases}$$

It is seen from the expression of y' that for $x < 0$, $y' < 0$, for $0 < x < 1$, $y' > 0$. By Theorem 4, the given function has a local minimum at the critical point: $x = 0$: $y_{\min}(0) = (0 - 3)^2 e^{\circ} = 9$.

As $f'(1) = 0$, $f'(3) = 0$, $x = 1$, and $x = 3$ are stationary points for the function $f(x)$. From the expressions of $f'(x)$ for the case $x > 0$ is seen that for $0 < x < 1$, $f'(x) > 0$, for $1 < x < 3$, for $f'(x) > 0$, for $x > 3$, $f'(x) > 0$. By Theorem 4, the function $f(x)$ has a local maximum at the point $x = 1$, a local minimum at the point $x = 3$: $y_{\max.}(1) = (1-3)^2 e^1 = 4e$; $y_{\min.}(3) = 0$.

Answer: 1) $y_{\min}\left(-\frac{1}{\sqrt{2}}\right)=-\frac{1}{2}$, $y_{\max}\left(\frac{1}{\sqrt{2}}\right)=\frac{1}{2}$;

2) $y_{\min}(e)=e$;

3) $y_{\min}\left(2k\pi+\frac{\pi}{3}\right)=2k\pi+\frac{\pi}{3}-\sqrt{3}$;

$y_{\max}\left(-\frac{\pi}{3}+2k\pi\right)=-\frac{\pi}{3}+2k\pi+\sqrt{3}$;

4) has no extremum; 5) has no extremum;

6) $y_{\min}(3)=0$; 8) has no extremum.

Problem 421. Study extremums of the following functions by means of higher order derivatives:

$$1)\, y=(x-3)^2;\ 2)\, y=\frac{1}{6}(x-1)^6+\frac{1}{2}(x-1)^4+(x-1)^2;$$
$$3)\, y=(x-1)^4.$$

Solution 3): Write the first-order derivative: $y'=4(x-1)^3$. Find the stationary point: $y'=0 \Rightarrow 4(x-1)^3=0 \Rightarrow x=1$. Write the second-order derivative: $y^{(2)}=12(x-1)^2$.

As $y^2(1) = 0$, we cannot study extremum by means of Theorem 5.

Write the third-order derivative: $y^{(3)} = 24(x - 1)$. As $y^{(3)}(1) = 0$, it is necessary to find the fourth-order derivative: As $y^{(4)} = 24 > 0$, $n = 4$ is an even number, by Theorem 6, $x = 1$ is a minimum point: $y_{\min}(1) = 0$.

Answer: 1) $y_{\min}(3) = 0$; 2) $y_{\min}(1) = 0$.

Problem 422. Find the largest and least values of the function on the indicated intervals:

$$1)\, y=x^5-5x^4+5x^3+1,\ x\in[-1;2];\ 2)\, y=\frac{x^4+1}{x^2+1},\ x\in[-1;1];$$
$$3)\, y=x-2\ln x,\ x\in\left[\frac{3}{2};e\right];\ 4)\, y=(x-3)e^{|x+1|},\ x\in[-2;4];$$
$$5)\, y=2\sin x+\sin 2x,\ x\in\left[0;\frac{3\pi}{2}\right];\ 6)\, y=\begin{cases}-x^2, & -1\le x\le 0,\\ 2ex\ln x, & 0<x\le 2.\end{cases}$$

Solution 4): At first find the critical points of the given function. This function is differentiable at all inner points of the interval [–2;4] except the point $x = -1$, and

$$y' = \begin{cases} e^{-x-1} - (x-3)e^{-x-1} = (4-x)e^{-x-1}, & -2 < x < -1, \\ e^{x+1} + (x-3)e^{x+1} = (x-2)e^{x+1}, & -1 < x < 4. \end{cases}$$

The function $y' = 0$ has a unique root: $x = 2$.

Thus, this function has two critical points: $x = 2$ is a stationary number, at the point $x = -1$ there is no derivative.

Find the values of the function at critical points:

$$y(-1) = (-1-3)e^{|-1+1|} = -4, \; y(2) = (2-3)e^{|2+1|} = -e^3.$$

Now find the values of the function on the edge points of the interval [– 2;4]:

$$y(-2) = (-2-3)e^{|-2+1|} = -5e, \; y(4) = (4-3)e^{|4+1|} = e^5.$$

As $-e^3$ is the least, e^5 is the largest number among the found numbers $-e^3$, $-5e$ -4, e^5,the least value of the given function on the interval [–2;4] is $-e^3$, the largest value is e^5.

Answer: 1) the largest value is 2, the least value is –10;

2) the largest value is 1, the least value is $2\sqrt{2}-2$;

3) the largest value is e –2, the least value is 2 – 2 ln 2;

5) the largest value is $\frac{3\sqrt{3}}{2}$, the least value is – 2;

6) the largest value is 4e ln 2, the least value is – 2.

Problem 423. Prove the inequalities:

$$1)\, e^x \geq 1+x;\; 2)\ln(1+x) > \frac{x}{x+1},\; (x>0);$$

$$3)\cos x \geq 1 - \frac{x^2}{2};\; 4)\sin x > \frac{2}{\pi}x,\; \left(0 \leq x \leq \frac{\pi}{2}\right);$$

$$5)\sin x + tg\, x > 2x,\; \left(0 < x < \frac{\pi}{2}\right).$$

Solution of 1): Denote $f(x) = e^x - 1 - x$, study the extremum of the function $f(x)$.

$$f'(x) = e^x - 1,\ f'(x) = 0 \Rightarrow e^x - 1 = 0 \Rightarrow x = 0.$$

As $f^{(2)}(x) = e^x \Rightarrow f^{(2)}(0) = 1 > 0$, by the second sufficient condition theorem on the existence of extremum, at the point $x = 0$ the function $f(x)$ has a local minimum. This minimum is the least value of this function, that is, $f(x) \geq f(0)$. Hence $f(0) = 0$, we get

$$f(x) \geq 0 \Rightarrow e^x - 1 - x \geq 0 \Rightarrow e^x \geq 1 + x.$$

Home tasks

Problem 424. Find the monotonicity intervals of the functions:

$$1)\ f(x) = 8x^3 - x^4;\ 2)\ f(x) = x^2 \ln x;$$

$$3)\ f(x) = \frac{1}{x} + \frac{2x}{x^2 - 1};\ 4)\ f(x) = \sqrt{2x^3 + 9x^2}.$$

Answer: 1) $(-\infty;6)$ is an increase, $(6;+\infty)$ is a decrease;

2) $\left(0;\frac{1}{\sqrt{e}}\right)$ is an increase, $\left(\frac{1}{\sqrt{e}};+\infty\right)$ is a decrease;

3) $(-\infty;-1)$, $(-1;0)$, $(0;1)$; $(1;+\infty)$ is a decrease;

4) $\left(-\frac{9}{2};-3\right)$, $(1;+\infty)$ is an increase, $(-3;0)$ is a decrease interval.

Problem 425. Find the extremums of the functions:

$$1)\ y = \frac{2x^2 - 1}{x^4};\ 2)\ y = x - 2\ln x;\ 3)\ y = e^x \cos x;$$

$$4)\ y = \frac{x}{x^2 + 4};\ 5)\ y = \sin^3 x + \cos^3 x;\ 6)\ y = \left(3 - x^2\right)e^x;$$

$$7)\ y = x^x;\ 8)\ f(x) = e^{|x+1|}.$$

Answer: 1) $y_{\max}(-1) = 1,\ y_{\max}(1) = 1$;

2) $y_{\min}(2) = 2(1 - \ln 2)$;

3) $y_{\max}\left(2k\pi+\frac{\pi}{4}\right)=e^{2k\pi+\frac{\pi}{4}}\cdot\frac{\sqrt{2}}{2}$,

$y_{\min}\left(2k\pi+\frac{5\pi}{4}\right)=-e^{2k\pi+\frac{\pi}{4}}\cdot\frac{\sqrt{2}}{2}$,

4) $y_{\max}(2)=\frac{1}{4}$, $y_{\min}(-2)=-\frac{1}{4}$;

5) $y_{\max}(2k\pi)=1$, $y_{\max}\left(2k\pi+\frac{\pi}{2}\right)=1$,

$y_{\max}\left(2k\pi+\frac{5\pi}{4}\right)=-\frac{\sqrt{2}}{2}$,

$y_{\min}(2k\pi+\pi)=-1$, $y_{\min}\left(2k\pi+\frac{3\pi}{2}\right)=-1$,

$y_{\min}\left(2k\pi+\frac{\pi}{4}\right)=\frac{\sqrt{2}}{2}$, k is an integer;

6) $y_{\min}(-3)=-6e^{-3}$, $y_{\max}(1)=2e$;

7) $y_{\min}\left(\frac{1}{e}\right)=\left(\frac{1}{e}\right)^{\frac{1}{e}}$; 8) $y_{\min}(-1)=1$.

Problem 426. Study the extremum of the function given in the implicit form:

$$x+y=xy(y-x),\quad |y|<|x|.$$

Answer: $y_{\max}(-1)=\sqrt{2}-1$, $y_{\min}(1)=1-\sqrt{2}$.

Problem 427. Study the extremum of the function given in parametric form:

$$x=\ln\sin\frac{t}{2},\qquad y=\ln\sin t.$$

Answer: $y_{\max}\left(-\frac{\ln 2}{2}\right)=0$.

Problem 428. Study the extremum of the function given in parametric form:

$$1)\ y=x^3-6x^2+9,\ x\in[-1;2];\ 2)\ y=x^2\ln x,\ x\in[1;e];$$

$$3)\ y=x+\sqrt{x},\ x\in[0;4];\ 4)\ y=\arccos x^2,\ x\in\left[-\frac{\sqrt{2}}{2};\frac{\sqrt{2}}{2}\right];$$

$$5)\ y=arctg\,x-\frac{1}{2}\ln x,\ x\in\left[\frac{1}{\sqrt{3}};\sqrt{3}\right];\ 6)\ y=(x-3)^2e^{|x|},\ x\in[-1;4];$$

$$7)\ f(x)=\begin{cases}2x^2+\dfrac{2}{x^2}, & -2\le x<0;\ 0<x\le 2,\\ 1, & x=0.\end{cases}$$

Answer: 1) 9 is the largest value, –7 is the least value;

2) e^2 is the largest value, 0 is the least value;

3) 6 is the largest value, 0 is the least value;

4) $\frac{\pi}{2}$ is the largest value, $\frac{\pi}{3}$ is the least value;

5) $\frac{\pi}{6}+0.25\ln 3$ is the largest value, $\frac{\pi}{3}-0.25\ln 3$ is the least value;

6) e^4 is the largest value, 0 is the least value;

7) has no largest value, 1 is the least value.

Problem 429. Prove the inequalities:

$$1)\ e^x\ge ex,\ x\in(-\infty;+\infty);\ 2)\ e^x>1+\ln(1+x);$$

$$3)\ tg\,x>x+\frac{x^3}{3},\ 0<x<\frac{\pi}{2};\ 4)\ arctg\,x\le x,\ x\ge 0.$$

4.2 DIRECTION AND TURNING POINTS OF THE CONVEXITY OF THE GRAPH OF A FUNCTION. ASYMPTOTES OF THE GRAPH OF A FUNCTION

Assume that the function $y=f(x)$ is a function determined on the interval $(a;\ b)$ and differentiable on this interval. Then this curve that is the graph of this function has a tangent at the point with any abscissa $x\in(a,\ b)$ and this tangent is not parallel to OY axis.

Definition. If the part of the graph of the function $y = f(x)$ that corresponds to the interval $(a; b)$ is arranged below (upper) of any tangent, then it is said that the graph of this function has a upwards (downwords) directed convexity on the interval (a, b).

Theorem 1. (A sufficient condition on direction of convexity of a curve). If the function $y = f(x)$ has a second-order continuous derivative on the interval (a, b) and $f^{(2)}(x) \geq 0$ ($f^{(2)}(x) \leq 0$), everywhere on this interval, then the graph of the function $f(x)$ has downwards (upwards) directed convexity on the interval $(a; b)$.

For example, when for the function $f(x) = x^3 - 3x^2 - 4$ for $x < 1, f^{(2)}(x) = 6(x-1)$, for $x > 1, f^{(2)}(x) < 0$, as $f^{(2)}(x) > 0$ by Theorem 1 the graph of the function $f(x)$ has upwards directed convexity on the interval $(-\infty; 1)$, and downwards directed convexity on the interval $(1; +\infty)$.

Definition. If the function $f(x)$ determined in certain neighbourhood of the point $x = c$, continuous at the point c and possessing continuous or discontinuous derivative at this point, changes the direction of its convexity when passing through the point c it is said that the graph of the function $f(x)$ has a turning at the point $M(c; f(c))$. The point $M(c; f(c))$ is called a turning point. When passing through the point $x = c$, the graph of the function changes the direction of convexity, but has no derivative at the point c, then $M(c; f(c))$ is called a nodal point.

Theorem 2. (Necessary condition for the existence of turning point). If $M(c; f(c))$ is a turning point for the function $y = f(x)$ that possesses the second-order continuous derivative at the point $x = c$, then $f^{(2)}(c) = 0$.

Theorem 3. (The first sufficient condition for the existence of a turning point). Assume that in the neighborhood of the point c (except the point c) the function $y = f(x)$ has a second-order derivative. If in this neighborhood, $f^{(2)}(x)$ has different signs in the left and right sides of the point c, then $M(c; f(c))$ is a turning point of the graph of the function $f(x)$ (if there is a second-order continuous derivative at the point c, in addition, it is assumed $f^{(2)}(c) = 0$).

Theorem 4. (The second sufficient condition for the existence of a turning point). If the function $y = f(x)$ has a third-order finite derivative at the point c and satisfies the conditions $f^{(2)}(c) = 0, f^{(3)}(c) \neq 0$, then the point $M(c; f(c))$ is a turning point of the graph of this function.

Definition. If even one of the limits $\lim_{x\to a+0} f(x)$ or $\lim_{x\to a-0} f(x)$ is $+\infty$ or $+\infty$, the straight line $x = a$ is the vertical asymptote of the graph of the function $y = f(x)$.

For example, for the function $f(x) = \frac{1}{x^2 - 4}$, as $\lim_{x\to 2} f(x) = \infty$, $\lim_{x\to -2} f(x) = \infty$, the straight lines $x = -2$ and $x = 2$ are vertical asymptotes of the graph of this function.

Definition. If at rather large values of the argument x it is possible to represent the function $f(x)$ in the form $f(x) = kx + b + \alpha(x)$, $\left(\lim_{x\to +\infty} \alpha(x) = 0\right)$, then the straight line $Y = k + b$ is called a sloping asymptote of the function $f(x)$.

Theorem 5. The necessary and sufficient condition for the straight line $Y = k + b$ a sloping asymptote of the graph of the function $y = f(x)$ is the existence of both of the limits

$$\lim_{x\to -+\infty} \frac{f(x)}{x} = k, \ \lim_{x\to -+\infty} \left(f(x) - kx\right) = b.$$

Problems to be solved in auditorium

Problem 430. Find the convexity intervals and turning points of the graph of the function:

$$1)\ y = x^7 + 7x + 1;\ 2)\ y = \frac{1}{12}x^4 - \frac{1}{6}x^3 - x^2 + 3;$$

$$3)\ y = x^3 \ln x + 1;\ 4)\ y = arctg\left(\frac{1}{x}\right);\ 5)\ y = 2 - \left|x^5 - 1\right|.$$

Solution of 2): At first find the first and second-order derivatives:

$$y' = \frac{1}{3}x^3 - \frac{1}{2}x^2 - 2x,\ y'' = x^2 - x - 2.$$

From the expression of y'' it is seen that for $x = -1$ and $x = 2$, $y'' = 0$,
for $x \in (-\infty; -1)$, $y'' > 0$,
for $x \in (-1;2)$, $y'' < 0$,

for $x \in (2; +\infty)$, $y'' < 0$. By Theorem 1, the graph of the given function has downwards directed convexity on the intervals $(-\infty; -1)$ and $(2; +\infty)$, the upwards directed convexity on the interval $(-1;2)$. Taking into account that

$$y(-1)=\frac{1}{12}+\frac{1}{6}-1+3=2.25,\ y(2)=\frac{4}{3}-\frac{4}{3}-4+3=-1.$$

By Theorem 3 we can state that (–1;2.25) and (2;–1) are turning points of the graph of the given function.

Answer: 1) Has upwards directed convexity on the interval $(-\infty; 0)$; downwards directed convexity on the interval $(0; +\infty)$, $M(0;1)$ is the turning point; 3) has upwards directed convexity on the interval $\left(0;e^{-\frac{5}{6}}\right)$; has downwards directed convexity on the interval $\left(e^{-\frac{5}{6}};+\infty\right)$, $M\left(e^{-\frac{5}{6}};1-\frac{5}{6}e^{-\frac{5}{6}}\right)$ is the turning point; 4) has upwards directed convexity on the interval $(-\infty; 0)$, has no turning point $(0; +\infty)$ (at the point $x = 0$ the function is discontinuous; 5) has upwards directed convexity on the intervals $(-\infty; 0)$ and $(0; +\infty)$, downwards directed convexity on the interval (0;1), $M(0;1)$ is a unique turning point $N(1;2)$, a nodal point.

Problem 431. Find the turning point of the graph of the function $y=f(x)$ given in parametric form:

$$1)\, x=te^{t},\ y=te^{-t},\ t>0;\ 2)\, x=\frac{2t^2+2}{t},\ y=\frac{t^3+3t+1}{t^2},\ 0<t<1.$$

Answer: $1)\left(\sqrt{2}e^{\sqrt{2}};\ \sqrt{2}e^{-\sqrt{2}}\right);\ 2)\left(5;\frac{21}{2}\right).$

Guideline: Calculate the derivatives by the formulas

$$y'_x=\frac{y'_t}{x'_t},\ y'_{xx}=\left(y'_x\right)'_t\cdot\frac{1}{x'_t}.$$

Problem 432*. Prove the inequalities:

$$1)\left(\frac{x+y}{2}\right)^n\le\frac{x^n+y^n}{2},\ x\ge 0, y\ge 0,\ n \text{ is a natural number;}$$

$$2)\, e^{\frac{x+y}{2}}\le\frac{e^x+e^y}{2},\quad x, y \text{ are any real numbers.}$$

Solution of 1): We can prove that if the graph of the function $f(x)$ has a downwards directed convexity on the interval $(a; b)$, then $\alpha_1 \ge 0$, $\alpha_2 \ge 0$, for $\alpha_1 + \alpha_2 = 1$, for any points x_1 and x_2 from the interval $(a; b)$ the function $f(x)$ will satisfy the condition

$$f(\alpha_1 x_1 + \alpha_2 x_2) \le \alpha_1 f(x_1) + \alpha_2 f(x_2). \qquad (*)$$

Therefore, sometimes the inequality (*) is accepted as definition of downwards directed convexity of the graph of the function.

On the semi-axis $(0; +\infty)$ let us consider the function $f(x) = x^n$. As $f''(x) = n(n-1)x^{n-2} \ge 0, x \in (0; +\infty)$, the graph of this function has a downwards directed convexity on the semi-axis $(0; +\infty)$. Then, if in (*) we take $f(x) = x^n$, we can write

$$(\alpha_1 x_1 + \alpha_2 x_2)^n \le \alpha_1 x_1^n + \alpha_2 x_2^n.$$

If in the last inequality we write $\alpha_1 = \alpha_2 = \frac{1}{2}$, $x_1 = x$, $x_2 = y$, we get the inequality that we want to prove.

Problem 433. Find the asymptotes of the graphs of the function:

$$1)\, y = \sqrt[5]{\frac{x}{x-2}};\ 2)\, y = \frac{\sqrt{|x^2-3|}}{x};$$
$$3)\, y = \frac{\ln(x+1)}{x^2} + 2x;\ 4)\, y = x\ln\left(e + \frac{1}{x}\right).$$

Solution of 3): The given function is everywhere continuous on the semi-axis $[-1; +\infty]$ except the points $x = 0$ and $x = -1$. Therefore, it may have two vertical asymptotes.

As $\lim\limits_{x\to 0\pm 0} y = \lim\limits_{x\to 0\pm 0}\left(\frac{1}{x}\cdot\frac{\ln(x+1)}{x} + 2x\right) = \pm\infty$,

$x = 0$ is a vertical asymptote.

As $\lim\limits_{x\to -1+0} y = \lim\limits_{x\to -1+0}\left(\frac{\ln(x+1)}{x^2} + 2x\right) = -\infty$, $x = -1$ is a vertical asymptote.

Now define if there exists a sloping asymptote. For that, by Theorem 5 at first we consider the limit $\lim\limits_{x\to\infty}\frac{y}{x}$:

$$\lim_{x\to\infty}\frac{y}{x} = \lim_{x\to\infty}\left(\frac{\ln(x+1)}{x^3} + 2\right) = \lim_{x\to\infty}\frac{\frac{1}{x+1}}{3x^2} + 2 =$$

$$= \lim_{x\to\infty} \frac{1}{3x^2(x+1)} + 2 = 0 + 2 = 2 .$$

Thus, if the graph of the given function has the sloping asymptote $Y = kx + b$, then $k = 2$. Again, by Theorem 5 we consider the limit

$$\lim_{x\to\infty}(y - kx) = \lim_{x\to\infty}(y - 2x),$$

$$\lim_{x\to\infty}(y - 2x) = \lim_{x\to\infty}\left(\frac{\ln(x+1)}{x^2} + 2x - 2x\right) = \lim_{x\to\infty}\frac{1}{x}\cdot\frac{\ln(x+1)}{x} = 0 .$$

So, $b = 0$. The graph of the function has the sloping asymptote $Y = 2x$.

Answer: 1) Vertical asymptote is $x = 2$, sloping asymptote is $y = 1$;

2) Vertical asymptote is $x = 0$, sloping asymptotes are $y = 1$ and $y = -1$;

4) Vertical asymptote is $x = -\frac{1}{e}$, sloping asymptote is $y = x + \frac{1}{e}$.

Home tasks

Problem 434. Find the convexity intervals and turning points of the graph of the function:

$$1)\, y = x^4 + 6x^2;\ 2)\, y = \sqrt[3]{(x-2)^5} + 3;$$
$$3)\, y = xe^{2x} + 1;\ 4)\, y = x\ln|x|.$$

Answer: 1) Everywhere the graph has downwards directed convexity; 2) on the interval $(-\infty; 2)$ has upwards directed, on the interval $(2; +\infty)$ has downwards directed convexity, $M(2;3)$ is a turning point; 3) on the interval $(-\infty; -1)$ has upwards directed, on the interval $(-1; +\infty)$ has downwards directed convexity, $M(-1; 1-e^{-2})$ is a turning point; 4) on the interval $(-\infty; 0)$ has upwards directed, on the interval $(0; +\infty)$ downwards directed convexities.

Problem 435*. Prove the inequality:

$$\frac{x\ln x + y\ln y}{x+y} > \ln\frac{x+y}{2},\ x > 0,\ y > 0 .$$

Guideline: Use that on the interval $(0; +\infty)$ the graph of the function $f(x) = x\ln x$ has downwards direct convexity, and formula (*).

Problem 436. Find the turning points of the graph of the function $y = f(x)$ given in parametric form:

$$x = \frac{t^2}{t-1},\ y = \frac{t^3}{t-1},\ t > 2.$$

Answer: $\left(\frac{9}{2};\frac{27}{2}\right)$.

Problem 437. Find the asymptotes of the graph of the function:

$$1)\ y = \sqrt[3]{x^3 - x^2};\ 2)\ y = 3x + arctg5x;\ 3)\ y = \frac{\sin x}{x};$$

$$4)\ y = x\,arc\sec x;\ 5)\ y = \frac{3x}{x-1} + 3x.$$

Answer: 1) has no vertical asymptote, sloping asymptote

$$y = x - \frac{1}{3};$$

2) has two sloping asymptotes:

$$y = 3x + \frac{\pi}{2} \text{ and } y = 3x - \frac{\pi}{2};$$

3) sloping asymptote $y = \frac{\pi}{2}x - 1$;

5) has no vertical asymptote, sloping asymptote $x = 1,\ y = 3x + 3$.

4.3 GRAPHING OF A FUNCTION

For graphing a function, it is appropriate to study it by the following scheme:

1) Finding the domain of definition of a function.
2) Determination of evenness, oddness, and periodicity of a function.
3) Finding the intersection points of the graph of the function and coordinate axes.
4) If it possible, to determine if a function is positive or negative.
5) Finding the discontinuity points.

6) Finding the asymptotes of a function (if it exists).
7) Calculation of unidirectional limits of domain of definition at boundary points and discontinuity points.
8) Finding increasing, decreasing intervals, and extremums of a function.
9) Determination of convexity of intervals and turning angles of the graph of a function.

Depending on the properties of the would-be constructed function, some of the above stages may be not studied; some of them should be studied more thoroughly. In some cases, for drawing more exact graph, the coordinates of some points arranged on the graph are also found.

Problems to be solved in auditorium

Problem 438. Study the functions and construct their graphs:

$$1)\, y=\frac{1}{4}x^2\left(x^2-3\right)^2;\ 2) y=\frac{1-x^3}{x^2};\ 3)\, y=\frac{x^3}{2(x-1)^2};$$

$$4)\, y=\sin x+\cos x;\ 5)\, y=x+2arc\,ctg\,x;\ 6)\, y=\sqrt[3]{\left|x^2-1\right|}.$$

Solution of 2): at first study the function $y=\frac{1-x^3}{x^2}$ by the above scheme.

1. The function was determined at all points with $x \neq 0$.
2. The function is not even, odd, or periodic.
3. Find the intersection points of the function and coordinate axes. As $y = 0$ at the intersection point with the axis OX, then $\frac{1-x^3}{x^2}=0 \Rightarrow x=1$, that is, the graph intersects the abscissa axis only at one point, (1;0). The graph does not intersect the ordinate axis.
4. It is seen from explicit expressions of the function that for $-\infty < x < 0$ and $0 < x < 1$, for $y > 0$, $1 < x < \infty$ for $y < 0$. This means that on the intervals $(-\infty; 0)$ and $(0;1)$ the graph of the function is above the abscissa axis, on the interval $(1; +\infty)$ below the abscissa axis.
5. At the point $x = 0$ the function has infinite discontinuity:

$$\lim_{x\to -0} y=+\infty,\ \lim_{x\to +0} y=+\infty$$

6. As the function has infinite discontinuity at the point $x = 0$, the straight line $x = 0$ (ordinate axis) is a vertical asymptote of the graphs of this function. Let us search sloping asymptotes of the function:

$$k = \lim_{x\to+\infty} \frac{y}{x} = \lim_{x\to+\infty} \frac{1-x^3}{x^3} = \lim_{x\to+\infty} \left(\frac{1}{x^3} - 1\right) = -1,$$

$$b = \lim_{x\to+\infty} (y - kx) = \lim_{x\to+\infty} \left(\frac{1-x^3}{x^2} + x\right) = \lim_{x\to+\infty} \frac{1}{x^2} = 0.$$

So, $y = kx + b \Rightarrow y = -x$ is the sloping asymptote of this function.

7. If we write this function in the form $y = \frac{1-x^3}{x^2} = \frac{1}{x^2} - x$, we get $\lim_{x\to-\infty} y = +\infty$, $\lim_{x\to+\infty} y = -\infty$. This means that moving to the left along the abscissa axis, the graph of the function ascends, moving to the right it descends.

8. Find the increase, decrease intervals, and extremum of the function. As $y' = -\frac{x^3+2}{x^3}$, $x = -\sqrt[3]{2}$ is a critical point $\left(y'\left(-\sqrt[3]{2}\right) = 0\right)$. The point $x = 0$ has no first-order derivative, but as at this point the function is discontinuous, the point $x = 0$ is not a critical point. As $y'' = \frac{6}{x^4}$, $y''\left(-\sqrt[3]{2}\right) > 0$, at the point $x = -\sqrt[3]{2}$ the function has a local minimum:

$$y_{\min} = y\left(-\sqrt[3]{2}\right) = \frac{3}{\sqrt[3]{4}}.$$

It is seen from the expression of y' that for $-\infty < x < -\sqrt[3]{2}$, or $0 < x < +\infty$, $y' < 0$ that is, the function decreases. For $-\sqrt[3]{2} < x < 0$, $y' > 0$, that is, the function increases.

9. It is seen from the expression $y'' = \frac{6}{x^4}$ that $y'' \neq 0$. y'' was not determined at the point $x = 0$. So, the graph of the function has no turning point. As at all the points of domain of definition $y'' > 0$, everywhere the graph of the function has downwards directed convexity. Giving the values $-2; -1; \frac{1}{2}; 2$ to x, we find that the

points $\left(-2;2\frac{1}{4}\right)$, $\left(-1;\frac{1}{2}\right)$, $\left(\frac{1}{2};3\frac{1}{2}\right)$, $\left(2;-1\frac{3}{4}\right)$ are also on the graph of this function.

According to our researches and to these points we construct the graph of the function $y=\frac{1-x^3}{x^2}$ (Fig. 4.1).

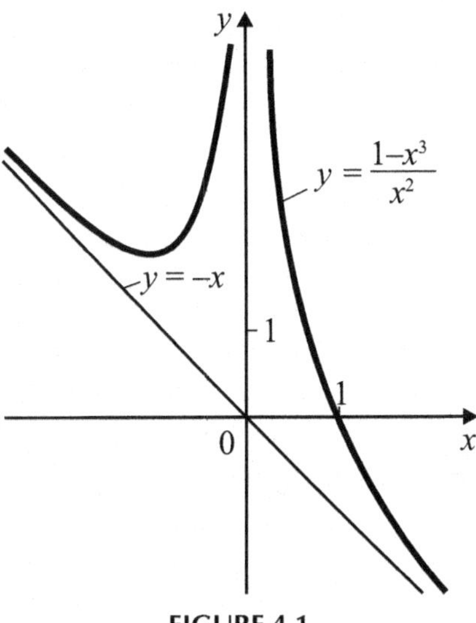

FIGURE 4.1

Solution of 5): At first study the function y = x + 2*arcctg x*.

1. The function was determined on the real axis.
2. The function is not odd, even, or periodic.
3. The graph intersects the axis *OY* at the point with the ordinate $y=\pi$:

$$y=0+2arc\,ctg\,0=2\cdot\frac{\pi}{2}=\pi\,.$$

4. It has no vertical asymptote. Find sloping asymptotes:

$$k=\lim_{x\to\pm\infty}\frac{y}{x}=\lim_{x\to\pm\infty}\left(1+\frac{2arc\,ctg\,x}{x}\right)=1,$$

$$b_1 = \lim_{x\to+\infty}(y-kx) = \lim_{x\to+\infty} 2arc\,ctg\,x = 2\cdot 0 = 0$$

$$b_2 = \lim_{x\to-\infty}(y-kx) = \lim_{x\to-\infty} 2arc\,ctg\,x = 2\cdot\pi = 2\pi .$$

Thus, the graph of the function has two sloping asymptotes as $y = x$ and $y = x + 2\pi$.

5. The derivative $y' = 1 - \frac{2}{1+x^2} = \frac{x^2-1}{x^2+1}$ vanishes at the points $x = \pm 1$.

As $y'' = \frac{4x}{\left(1+x^2\right)^2}$, $y''(-1) < 0$, $y''(1) > 0$ this function has local maximum $\left(y_{\max} = y(-1) = \frac{3\pi}{2} - 1\right)$ at the point $x = -1$, and local minimum $\left(y_{\min} = y(1) = \frac{\pi}{2} + 1\right)$ at the point $x = 1$. It is clear from the expression $y' = \frac{x^2-1}{x^2+1}$ that for $-\infty < x < -1$ or $1 < x < +\infty$, $y' > 0$, that is, on the intervals $(-\infty; -1)$ and $(1; +\infty)$ the function increases, on $(-1;1)$ decreases. The graphs of the studied functions have been described in Figure 4.2.

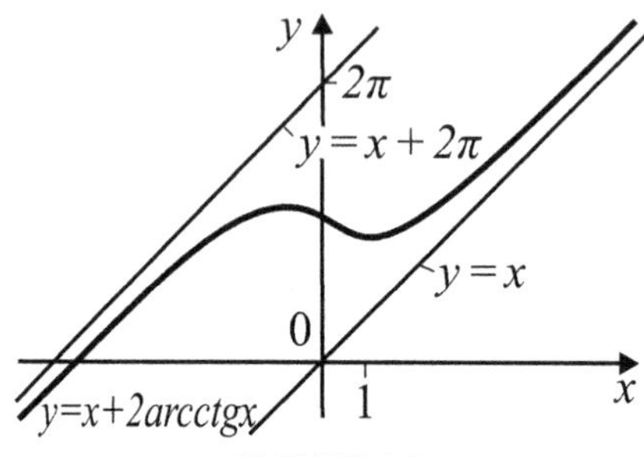

FIGURE 4.2

6. It is seen from the formula $y'' = \frac{4x}{\left(1+x^2\right)^2}$ that $y''(0) = 0$, for $x < 0$, $y'' < 0$, for $x > 0$ $y'' > 0$. Thus, on the interval $(-\infty; 0)$ the graph of the function has upwards directed convexity, on the interval $(0; +\infty)$ the downwards directed convexity. Based on these studies we construct the graph of the function

$$y = x + 2arc\,ctg\,x .$$

Home tasks

Problem 439. Study the functions and construct their graphs:

$$1)\, y = \frac{1}{6}x^3\left(x^2 - 5\right);\; 2)\, y = \frac{x^4}{x^3 - 1};\; 3)\, y = \sqrt{\left|x^2 - 2\right|^3};$$

$$4)\, y = \frac{1}{\sin x + \cos x};\; 5)\, y = x^2 \ln^2 x.$$

KEYWORDS

- **monotone function**
- **local extremum**
- **downwards convexity**
- **upwards convexity**
- **turning points**
- **vertical asymptote**
- **horizontal asymptote**

CHAPTER 5

HIGHER ALGEBRA ELEMENTS

CONTENTS

ABSTRACT

In this chapter, we give brief theoretical materials on complex numbers, operations on them, polynomials dependent on one variable, simple and repeated roots of a polynomial, the Horner system, and 21 problems.

5.1 COMPLEX NUMBERS AND OPERATIONS

When x, y are any real numbers, we consider the set of all possible numbers in the form (x, y). In this set, every pair (x, y) is considered as one element. When the sum and product of any two elements $z_1 = (x_1, y_1)$ and $z_2 = (x_2, y_2)$ are determined by the formulas

$$z_1+z_2 = (x_1, y_1) + (x_2, y_2) = (x_1 + x_2, y_1 + y_2),$$

$$z_1z_2 = (x_1, y_1)\cdot(x_2, y_2) = (x_1x_2-y_1y_2, x_1y_2+x_2y_1),$$

respectively, this set is said to be the set of complex numbers, its every element is called a complex number.

Two complex numbers $z_1 = (x_1, y_1)$ and $z_2 = (x_2, y_2)$ are considered to be equal for $x_1 = x_2, y_1 = y_2$.

x is called a real part of the complex number $z = (x, y)$ and is denoted as Re$z = x$. y is called an imaginary part of this complex number and is denoted as $Jmz = y$.

The complex numbers in the form $(x, 0)$ whose imaginary part equals zero are written as $(x, 0) = x$, and considered as a real number x. For the complex number $(0, 1)$ a new designation was accepted: $i = (0, 1)$.

i is called an imaginary unit. By the rule of multiplication of complex numbers

$$i^2 = i \cdot i = (0. 1) \cdot (0. 1) = (0\cdot 0 - 1\cdot1, 0\cdot1 + 0\cdot1) = (-1, 0) = -1.$$

By the rule of addition and multiplication of complex numbers, we can write any complex number $z = (x, y)$ in the form

$$z = (x, y) = (x, 0) + (0, 1)\cdot(y, 0) = x + iy \tag{5.1}$$

The notation of the complex number in the form $z = x + iy$ is called its algebraic form.

In the rectangular coordinate system to each complex number $z = (x, y) = x + iy$ we can associate one $M(x; y)$ point with abscissa x and ordinate y. In majority of cases, the vector $\overrightarrow{OM}$ is associated to the complex number $z = (x, y)$.

The number $r = \left|\overrightarrow{OM}\right| = \sqrt{x^2 + y^2}$ is called a modulus of the complex number, every solution of the equation

$$\cos\varphi = \frac{x}{\sqrt{x^2 + y^2}}, \sin\varphi = \frac{y}{\sqrt{x^2 + y^2}} \tag{5.2}$$

is called an argument of the complex number. Very often, the solution satisfying the condition $|\varphi| \le \pi$ is taken as an argument. As from (5.2) we get $x = r\cos\varphi$, $y = r\sin\varphi$, we can write the complex number $z = x + iy$ in the form

$$z = r(\cos\varphi + i\sin\varphi). \tag{5.3}$$

Equation 5.3 is called the notation of a complex number in trigonometric form.

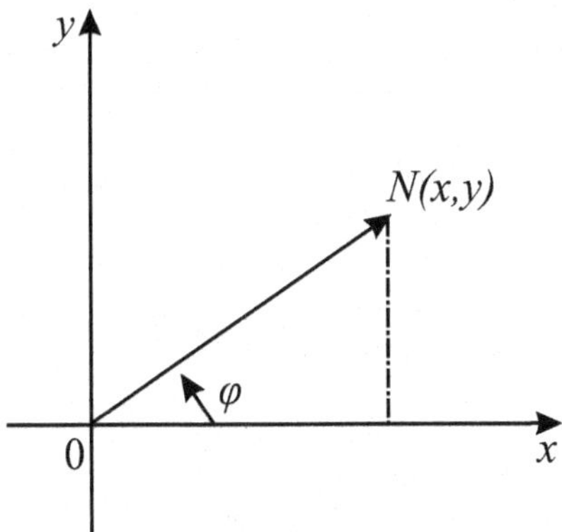

FIGURE 5.1

Sometimes the plane on which complex members are described, is called a complex plane, OX is called a real axis, OY an imaginary axis (Fig. 5.1).

We can take the arguments of positive real numbers as 0°, the argument of real negative numbers as π. Therefore, for example, we can write the number 5 in trigonometric form as $5 = 5(\cos 0° + i\sin 0°)$, the number –2 as $-2 = 2(\cos\pi + i\sin\pi)$.

Complex numbers with a zero real part are called absolutely imaginary numbers.

For $y > 0$, we can represent the argument of absolutely imaginary complex number iy as $\frac{\pi}{2}$, for $y < 0$ as $-\frac{\pi}{2}$. For example, we can write

$$3i = 3\left(\cos\frac{\pi}{2} + i\sin\frac{\pi}{2}\right), -5i = 5\left(\cos\left(-\frac{\pi}{2}\right) + i\sin\left(-\frac{\pi}{2}\right)\right).$$

The complex number $x - iy$ is called a complex number conjugated to the complex number $z = x + iy$ and is denoted as $\bar{z} = x - iy$.

For multiplication, division, and raising to power of complex numbers in the trigonometric form, the following formulas are true:

$$z_1 = r_1(\cos\varphi_1 + i\sin\varphi_1), z_2 = r_2(\cos\varphi_2 + i\sin\varphi_2),$$

$$z_1 \cdot z_2 = r_1 r_2(\cos(\varphi_1 + \varphi_2) + i\sin(\varphi_1 + \varphi_2)), \tag{5.4}$$

$$\frac{z_1}{z_2} = \frac{r_1}{r_2}(\cos(\varphi_1 - \varphi_2) + i\sin(\varphi_1 - \varphi_2)), \tag{5.5}$$

$$z^n = (r(\cos\varphi + i\sin\varphi))^n = r^n(\cos n\varphi + i\sin n\varphi). \tag{5.6}$$

n-th degree root of every complex number $z = r(\cos\varphi + i\sin\varphi)$ gives n number different complex numbers and these numbers may be found by the formula

$$\sqrt[n]{r(\cos\varphi + i\sin\varphi)} = \sqrt[n]{r}\left(\cos\frac{\varphi + 2k\pi}{n} + i\sin\frac{\varphi + 2k\pi}{n}\right), \quad k = 0, 1, \ldots, n-1. \tag{5.7}$$

The equality $e^{i\varphi} = \cos\varphi + i\sin\varphi$ is called the Euler formula.

Using the Euler formula, we can write (5.3) as follows

$$z = r(\cos\varphi + i\sin\varphi) = re^{i\varphi}. \tag{5.8}$$

The right side of (5.8) is called an exponential form of the complex number.

Problems to be solved in auditorium

Problem 440. Execute the operations:

$$1)\ (2+3i)(3-2i);\ 2)\ (1+i)^3;\ 3)\ \frac{2i}{1+i};\ 4)\ \frac{2+i}{3-i}.$$

Solution of 3): Method I. Denotes $\frac{2i}{1+i}=x+iy$. Then from the condition of equality of two complex numbers

$$2i=(1+i)(x+iy)\Rightarrow 2i=x+iy+ix+i^2y=x+i(x+y)-y=$$
$$=(x-y)+i(x+y)\Rightarrow 2i=(x-y)+i(x+y).$$

we get $\begin{cases}x-y=0,\\x+y=2\end{cases}\Rightarrow\begin{cases}2x=2\\x+y=2\end{cases}\Rightarrow\begin{cases}x=1,\\y=1\end{cases}$. So, $\frac{2i}{1+i}=1+i.$

Method II. Multiply the denominator and numeral of the fraction $\frac{2i}{1+i}$ by the conjugate of the denominator:

$$\frac{2i(1-i)}{(1+i)(1-i)}=\frac{2i-2i^2}{1-i^2}=\frac{2i+2}{1+1}=\frac{2(i+1)}{2}=i+1.$$

Answer: 1) $12+5i$; 2) $-2+2i$; 4) $\frac{1}{2}+\frac{1}{2}i.$

Problem 441. Write the following complex numbers in trigonometric form:

$$1)\ z=3;\ 2)\ z=-1;\ 3)\ z=3i;\ 4)\ z=-2i;$$
$$5)\ z=2-2i;\ 6)\ z=-\sqrt{3}-i;\ 7)\ z=-\sqrt{3}+i.$$

Solution of 6): As $z=-\sqrt{3}-i\Rightarrow x=\mathrm{Re}\,z=-\sqrt{3}$, $y = Jm\ z = -1$ for $r=\sqrt{x^2+y^2}=\sqrt{3+1}=2$. By the (5.2), we can write $\cos\varphi=\frac{x}{r}=\frac{-\sqrt{3}}{2}=-\frac{\sqrt{3}}{2}$, $\sin\varphi=-\frac{1}{2}$. It is clear from the last two equations that the angle φ falls within the third coordinate quarter: $\varphi=-\left(\frac{\pi}{2}+\frac{\pi}{3}\right)=-\frac{5\pi}{6}$. Writing the values of r and φ in (5.3), we get $-\sqrt{3}-i=2\left(\cos\left(-\frac{5\pi}{6}\right)+i\sin\left(-\frac{5\pi}{6}\right)\right)$.

Answer:

1) $3 = 3(\cos 0° + i \sin 0°$; 2) $-1 = 1(\cos \pi + i \sin \pi)$;

3) $3i = 3\left(\cos\frac{\pi}{2} + i\sin\frac{\pi}{2}\right)$; 4) $-2i = 2\left(\cos\left(-\frac{\pi}{2}\right) + i\sin\left(-\frac{\pi}{2}\right)\right)$;

5) $2-2i = 2\sqrt{2}\,(\cos(-45°) + i\sin(-45°)$; 7) $2\left(\cos\frac{5\pi}{6} + i\sin\frac{5\pi}{6}\right)$.

Problem 442. Calculate:

$$1)\ (1-i\sqrt{3})^6;\ 2)\ (-1+i)^5;\ 3)\ (-\sqrt{3}+i)^6.$$

Answer: 1) 64; 2) $4(1-i)$; 3) –64.

Guideline: Write the numbers in trigonometric form, and use (5.6).

Problem 443. Find: 1) $\sqrt[3]{1}$; 2) $\sqrt[6]{-1}$; 3) $\sqrt{i}$; 4) $\sqrt[3]{-1+i}$.

Solution of 1): Write the number under the radical sign, and use (5.7):

$$\sqrt[3]{1} = \sqrt[3]{\cos{}^{\circ} + i\sin 0°} = \cos\frac{0° + 360°\cdot k}{3} + i\sin\frac{0° + 360°\cdot k}{3} =$$
$$= \cos(120°k) + i\sin(120°k);\ k = 0,1,2.$$

$$k = 0 \text{ for } \left(\sqrt[3]{1}\right)_1 = \cos 0° + i\sin 0° = 1,$$

$$k = 1 \text{ for } \left(\sqrt[3]{1}\right)_2 = \cos 120° + i\sin 120° = -\frac{1}{2} + i\frac{\sqrt{3}}{2},$$

$$k = 2 \text{ for } \left(\sqrt[3]{1}\right)_3 = \cos 240° + i\sin 240° = -\frac{1}{2} - i\frac{\sqrt{3}}{2}.$$

Answer: 2) $\pm i,\ \pm\frac{\sqrt{3}}{2} \pm \frac{1}{2}i$; 3) $\pm\left(\frac{1}{\sqrt{2}} + \frac{1}{\sqrt{2}}i\right)$;

4) $\sqrt[6]{2}\,(\cos(45° + 120°\cdot k) + i\sin(45° + 120°\cdot k))$; $k = 0,1,2$.

Problem 444. Solve the equation: $x^4 + 4 = 0$.

Answer: $1+i,\ 1-i,\ -1+i, -1-i$.

Home tasks

Problem 445. Execute the operations:

$$1)\ (a+bi)(a-bi);\ 2)\ (3-2i)^2;\ 3)\ \frac{1+i}{1-i};\ 4)\ \frac{3+4i}{2i}.$$

Answer: 1) a^2+b^2; 2) $5-12i$; 3) i; 4) $2-\frac{3}{2}i$.

Problem 446. Decompose polynomials into linear co-factors:

$$1)\ x^2+1;\ 2)\ x^4-1;\ 3)\ x^2-2x+5.$$

Answer: 1) $(x+i)(x-i)$; 2) $(x+i)(x-i)(x+1)(x-1)$;
3) $(x-1+2i)(x-1-2i)$.

Problem 447. Write the complex numbers in trigonometric form:

$$1)\ z=3;\ 2)\ z=-2;\ 3)\ z=-\frac{1}{2}i;\ 4)\ z=5i;$$
$$5)\ z=1+i;\ 6)\ z=-1+i;\ 7)\ z=1-i;\ 8)\ z=-1-i.$$

Answer: 1) $3(\cos 0°+i\sin 0°)$; 2) $2(\cos\pi+i\sin\pi)$;

3) $\frac{1}{2}\left(\cos\left(-\frac{\pi}{2}\right)+i\sin\left(-\frac{\pi}{2}\right)\right)$; 4) $5\left(\cos\frac{\pi}{2}+i\sin\frac{\pi}{2}\right)$;

5) $\sqrt{2}\left(\cos\frac{\pi}{4}+i\sin\frac{\pi}{4}\right)$; 6) $\sqrt{2}\left(\cos\frac{3\pi}{4}+i\sin\frac{3\pi}{4}\right)$;

7) $\sqrt{2}\left(\cos\left(-\frac{\pi}{4}\right)+i\sin\left(-\frac{\pi}{4}\right)\right)$; 8) $\sqrt{2}\left(\cos\left(-\frac{3\pi}{4}\right)+i\sin\left(-\frac{3\pi}{4}\right)\right)$.

Problem 448. Calculate: 1) $\sqrt[3]{i}$; 2) $\sqrt[3]{-2+2i}$; 3) $\sqrt[3]{-1+i}$; 4) $\sqrt[4]{-8+8\sqrt{3}i}$.

Answer: 1) $\pm\frac{\sqrt{3}}{2}+\frac{1}{2}i,\ -i$; 2) $1+i,\ \sqrt[6]{8}(\cos 165°+i\sin 165°)$,
$\sqrt[6]{8}(\cos 285°+i\sin 285°)$; 3) $\sqrt[6]{2}(\cos\varphi+i\sin\varphi)$, $\varphi=45°,165°,285°$;
4) $\sqrt{3}+i,\ -1+\sqrt{3}i,\ -\sqrt{3}-i,\ 1-\sqrt{3}i.$

Problem 449. Solve the equation: $x^3 + 8 = 0$.

Answer: $-2,\ 1 \pm i\sqrt{3}.$

5.2 POLYNOMIALS DEPENDENT ON ONE VARIABLE

When $a_0, a_1, a_2, \ldots, a_n$ are known numbers, the expression

$$f(x) = a_0 x^n + a_1 x^{n-1} + a_2 x^{n-2} + \ldots + a_{n-1} x + a_n \tag{5.9}$$

is said to be a polynomial, every $a_i x^{n-i}$, $i = 0, 1, \ldots, n$ summand at the right side of (5.9) is called the term of the polynomial. Degree of highest order term of the polynomial is said to be degree of this polynomial. It is clear that for $a_0 \neq 0$, the degree of $f(x)$ equals n. In (5.9), instead of x we write a certain number c, the obtained number $f(c)$ is said to be the value of the polynomial at the point c. When the value of the polynomial at the point c equals zero, the number c is said to be the root of this polynomial. For example, as for the polynomial $f(x) = x^4 - x^3 + 2x^2 - x - 1$, $f(1) = 0$, number 1 is the root of this polynomial. For any two polynomials $f(x)$ and $g(x)$ one can always find polynomials $q(x)$ and $r(x)$ with degree less than the degree of $g(x)$ such that

$$f(x) = g(x)q(x) + r(x)\cdot \tag{5.10}$$

Equation 5.10 expresses the theorem on division with remainder. The following statement on division of polynomial into binomial, is true.

Theorem (Bezout). Necessary and sufficient condition for division of the polynomial $f(x)$ into the binomial $x - c$ is that the number c is the root of the polynomial $f(x)$.

The remainder obtained after dividing the polynomial $f(x)$ into the binomial $x - c$ equals $f(c)$, that is,

$$f(x) = (x - c)q(x) + f(c). \tag{5.11}$$

The following theorem resolves the problem on the existence of a root of each polynomial.

***Theorem* (the main theorem of algebra).** Every polynomial with degree not less than unit has at least one root.

Based on the main theorem of algebra and Bezout theorem, we can show that every polynomial with degree n ($n \geq 1$) has rightly n number roots. If the numbers $c_1, c_2, \ldots, c_n$ are the roots of an n-th degree polynomial with higher term coefficient a_0, then the equality

$$f(x) = a_0(x - c_1)(x - c_2) \cdot \ldots \cdot (x - c_n) \tag{5.12}$$

is true.

If for the polynomial $f(x)$ and the number c the equality

$$f(x) = (x - c)^k \varphi(x),\ \varphi(c) \neq 0 \tag{5.13}$$

is satisfied, c is called k times repeated root of $f(x)$. Here k is a natural number satisfying the condition $k \leq n$, $\varphi(x)$ is the polynomial with degree not exceeding the degree of $f(x)$.

Once repeated, root is called a simple root. When the complex number $\alpha = c + di$ is the root of real coefficient polynomial $f(x)$, the complex number $\bar{\alpha} = c - di$ adjoint to α is the root of $f(x)$.

It is known from the theorem on division with remainder that when dividing the n degree polynomial $f(x)$ into the binomial $x - c$, the degree of quotient polynomial $q(x)$ will be $n - 1$. If we denote the coefficients of $q(x)$ by $b_0, b_1, \ldots, b_{n-1}$, the remainder by r, we can find them by the following table called the Horner's scheme:

a_0	a_1	a_2	…	a_{n-1}	a_n
$b_0 = a_0$	$b_1 = cb_0 + a_1$	$b_2 = cb_1 + a_2$	…	$b_{n-1} = cb_{n-2} + a_{n-1}$	$r = cb_{n-1} + a_n$.

The rule for using the table: we take $b_0 = a_0$, for finding the next element of every subsequent column of the second row, we multiply c by the antecedent column element of the second row and add to the same column element of the first row.

The Horner scheme is used for finding the value of a polynomial at certain point and to determine how many times the root of a polynomial is repeated.

Problems to be solved in auditorium

Problem 450. Divide the polynomial $f(x)$ into the polynomial $g(x)$ with remainder:

1) $f(x) = 3x^5 + 7x^4+12x^3 + 17x^2 + 7x + 2$, $g(x) = x^3 + 2x^2 + 3x + 4$;
2) $f(x) = x^5 + 3x^4 - 2x^3 - 2x^2 + 5x + 1$, $g(x) = x^2 + 3x - 1$.

Solution of 2):

$$\begin{array}{l|l}
x^5 + 3x^4 - 2x^3 - 2x^2 + 5x + 1 & x^2 + 3x - 1 \\ \hline
x^5 + 3x^4 - x^3 & x^3 - x + 1 \\ \hline
\quad -x^3 - 2x^2 + 5x + 1 & \\
\quad -x^3 - 3x^2 + x & \\ \hline
\qquad x^2 + 4x + 1 & \\
\qquad x^2 + 3x - 1 & \\ \hline
\qquad\quad x + 2 &
\end{array}$$

$f(x) = g(x)\,(x^3 - x +1) + (x + 2)$.

Answer: 1) $f(x) = g(x)\,(3x^2+x+1) -2$.

Problem 451. Using the Horner scheme, divide the polynomial $f(x)$ into the binomial $g(x)$ with remainder:

$$1)\ f(x) = 2x^5 + 7x^4 - 8x^2 + 3x - 5,\ g(x) = x + 2;$$
$$2)\ f(x) = x^4 + 5x^3 - 6x^2 + 8x - 3,\ g(x) = x - 2.$$

Solution of 1): As $x + 2 = x - (-2)$ in the Horner scheme we take $c = -2$, $a_0 = 2$ and get:

2	7	0	−8	3	−5
2	(−2)·2+7=3	(−2)·3+0=−6	(−2)(−6)−8=4	(−2) ·4+3=−5	(−2)·(−5)−5=5.

So, $q(x) = 2x^4 + 3x^3 - 6x^2 + 4x - 5$, $r = 5$.

Therefore $2x^5 + 7x^4 - 8x^2 + 3x - 5 = (x+2)(2x^4 + 3x^3 - 6x^2 + 4x - 5) + 5$.

Answer:

$$2)\ x^4 + 5x^3 - 6x^2 + 8x - 3 = (x-2)(x^3 + 7x^2 + 8x + 24) + 45.$$

Problem 452. Determine how many times the root of the polynomial x_0 of the number x_0 is repeated:

$$1)\ f(x)=x^4-6x^3+10x^2-6x+9,\ x_0=3;$$
$$2)\ f(x)=x^5+6x^4+11x^3+2x^2-12x-8,\ x_0=-2.$$

Solution of 1): By the Horner scheme, at first we divide $f(x)$ into $(x-x_0)-a$ and then divide the obtained quotient polynomial again into $(x-x_0)-a$ and continue the process until we get a nonzero remainder. The root x_0 is repeated as many as the number of zeros.

	1	–6	10	–6	9
3	1	–3	1	–3	0
	1	0	1	0	
3	1	3	10		

As 2 zero remainders are obtained, $x_0=3$ is a twice-repeated root.

Answer: 2) three times repeated root.

Problem 453. Construct the least degree real coefficient polynomial with the given roots:

1) 2, 3, $1+i$ is a simple root, 1 is a twice-repeated root;

2) $2-3i$ is a three times repeated root;

3) i is twice repeated, -1 is a simple root.

Answer:

$$1)\ c(x-1)^2(x-2)(x-3)(x^2-2x+2)=$$
$$=c(x^6-9x^5+33x^4-65x^3+74x^2-46x+12);$$
$$2)\ c(x^2-4x+13)^3;$$
$$3)\ c(x^2+1)^2(x^2+2x-2)=$$
$$=c(x^6+2x^5+4x^4+4x^3+5x^2+2x+2),$$

$c\neq 0$ are arbitrary real numbers.

Problem 454. Knowing that the number i is the root of the polynomial $f(x)=x^4-5x^3+7x^2-5x+6$, decompose this polynomial into linear cofactors.

Answer: $(x^2+1)(x-2)(x-3)$.

Guideline: Taking into account that the number i is a root, divide the given polynomial into the polynomial $x^2 + 1$.

Problem 455. Knowing that the number $1+i$ is the root of the equation $x^4 - x^3 - 2x^2 + 6x - 4 = 0$, find the other roots of this equation.

Answer: $1-i$, -2, 1.

Home tasks

Problem 456. Divide the polynomial $f(x)$ into the polynomial $g(x)$ with remainder:

$$1)\ f(x) = x^5 - 3x^4 - 12x^3 + 21x^2 - 22x + 7,\ g(x) = x^3 + 2x^2 - 3x + 4;$$
$$2)\ f(x) = x^5 - 3x^4 + 3x^3 - 7x - 2,\ g(x) = x^3 - x^2 + 2x + 3.$$

Answer:
$$1)\ f(x) = g(x)(x^2 - 5x + 1) + (x + 3);$$
$$2)\ f(x) = g(x)(x^2 - 2x - 1) + (x + 1).$$

Problem 457. Using the Horner scheme, divide the polynomial $f(x)$ into the binomial $g(x)$:

$$1)\ f(x) = 2x^4 + 3x^3 + 4x - 3,\ g(x) = x + 2;$$
$$2)\ f(x) = 5x^5 - 2x^4 - 38x^3 - 4x + 1,\ g(x) = x - 3.$$

Answer:
$$1)\ f(x) = g(x)(2x^3 - x^2 + 2x) - 3;$$
$$2)\ f(x) = g(x)(5x^4 + 13x^3 + x^2 + 3x + 5) + 16.$$

Problem 458. Determine how many times the root of the polynomial $f(x)$ of the number x_0 is repeated:

$$1)\ f(x) = x^5 - 5x^4 + 7x^3 - 2x^2 + 4x - 8,\ x_0 = 2;$$
$$2)\ f(x) = x^5 + 7x^4 + 16x^3 + 8x^2 - 16x - 16,\ x_0 = -2.$$

Answer: 1) is a three times repeated root; 2) is a four times repeated root.

Problem 459. Construct the least degree real coefficient polynomial with the given roots:

1) $2+i$ is a simple root, 1 is a twice-repeated root: 2) $2-i$ is a twice-repeated root, -3 is a simple root.

Answer:

$$1)\ c(x^2-4x+5)(x-1)^2 = c(x^4-6x^3+4x^2+6x-5);$$
$$2)\ c(x^2-4x+5)^2(x+3),$$

$c \neq 0$ is an arbitrary real number.

Problem 460. Knowing that the number x_0 is the root of the polynomial $f(x)$, find the remaining roots of $f(x)$:

1) $f(x) = x^5 - 3x^4 + 4x^3 - 4x^2 + 3x - 1,\ x_0 = 1$ is a three times repeated root;

2) $f(x) = x^6 + 3x^5 + 4x^4 + 6x^3 + 5x^2 + 3x + 2,\ x_0 = i2$ is a twice-repeated root.

Answer: 1) i and $-i$; 2) $-i$ is a two times repeated root, -1 and -2 are simple roots.

KEYWORDS

- **imaginary unit**
- **complex number**
- **adjoint complex number**
- **modulus**
- **argument**
- **simple root**
- **repeated root**

INDEX

A

M

O

P

Printed in the United States
by Baker & Taylor Publisher Services